P9-CKX-899

Rediscovering Geography

New Relevance for Science and Society

Rediscovering Geography Committee
Board on Earth Sciences and Resources
Commission on Geosciences, Environment, and Resources
National Research Council

NATIONAL ACADEMY PRESS
Washington, D.C. 1997

titution Avenue, N.W. • Washington, DC 20418

his report was approved by the Governing Board of the are drawn from the councils of the National Academy ieering, and the Institute of Medicine. The members of the committee responsible for the report were chosen for their special competences and with regard for appropriate balance.

This report has been reviewed by a group other than the authors according to procedures approved by a Report Review Committee consisting of members of the National Academy of Sciences, the National Academy of Engineering, and the Institute of Medicine.

Support for this study was provided by the Association of American Geographers, Environmental Systems Research Institute, National Geographic Society, National Science Foundation (NSF Grant No. SBR-9319015/R), U.S. Bureau of the Census, U.S. Department of Transportation, U.S. Environmental Protection Agency, U.S. Geological Survey. Any opinions, findings, and conclusions or recommendations expressed in this material are those of the authors and do not necessarily reflect the views of the National Science Foundation.

Rediscovering Geography: New Relevance for Science and Society is available from the National Academy Press, 2101 Constitution Ave., NW, Box 285, Washington, DC 20055 (1-800-624-6242; http://www.nap.edu).

Library of Congress Cataloging-in-Publication Data
Rediscovering geography : new relevance for science and society / Rediscovering Geography Committee, Board on Earth Sciences and Resources, Commission on Geosciences, Environment, and Resources, National Research Council.
p. cm.
Includes bibliographical references and index.
ISBN 0-309-05199-1
1. Geography. I. National Research Council (U.S.). Rediscovering Geography Committee.
G116.R43 1997
910—dc21 97-4630

Cover art by Y. David Chung and James Lee.

The cover art represents the transformation from the traditional view of geography as a fact-based discipline to the one focused on the concerns of society and the scientific enterprise at large. The top image of an historical sea chart represents geography's long history and its roots in the discovery of facts about places. The middle image of a landscape and the bottom image of a crowded city scene represent the subject matter of modern geography: the environment and human society. The arrows represent geography's concerns with place and the flows of processes and phenomena between places. The dynamic globe represents the tools and techniques that have been developed by geographers and that are now being used in science, education, business, and government: geographic information systems, spatial analysis, and geographic visualization.

The crowded city scene is a detail from Mr. Chung's mural ''Metropolitan Scene.''

Y. David Chung is a graduate of the Corcoran School of Art and has exhibited widely throughout the country, including the Boston Museum of Fine Arts and the Whitney Museum of American Art.

Printed in the United States of America

REDISCOVERING GEOGRAPHY COMMITTEE

THOMAS J. WILBANKS, *Chair*, Oak Ridge National Laboratory, Tennessee
ROBERT McC. ADAMS, Smithsonian Institution, *Emeritus*, Basalt, Colorado
MARTHA E. CHURCH, Hood College, *Emerita*, Frederick, Maryland
WILLIAM A.V. CLARK, University of California, Los Angeles
ANTHONY R. DE SOUZA, Southwest Texas State University, San Marcos
PATRICIA P. GILMARTIN, University of South Carolina, Columbia
WILLIAM L. GRAF, Arizona State University, Tempe
JAMES W. HARRINGTON, George Mason University, Fairfax, Virginia
SALLY P. HORN, University of Tennessee, Knoxville
ROBERT W. KATES, Independent Scholar, Trenton, Maine
ALAN MACEACHREN, The Pennsylvania State University, University Park
ALEXANDER B. MURPHY, University of Oregon, Eugene
GERARD RUSHTON, University of Iowa, Iowa City
ERIC S. SHEPPARD, University of Minnesota, Minneapolis
BILLIE LEE TURNER II, Clark University, Worcester, Massachusetts
CORT J. WILLMOTT, University of Delaware, Newark

Staff

KEVIN D. CROWLEY, Study Director
JENNIFER T. ESTEP, Administrative Assistant
SHELLEY A. MYERS, Project Assistant

[1] Through February 1995.

COMMISSION ON GEOSCIENCES, ENVIRONMENT, AND RESOURCES

M. GORDON WOLMAN, *Chair*, The Johns Hopkins University, Baltimore, Maryland
PATRICK R. ATKINS, Aluminum Company of America, Pittsburgh, Pennsylvania
JAMES P. BRUCE, Canadian Climate Program Board, Ottawa, Ontario
WILLIAM L. FISHER, University of Texas at Austin
JERRY F. FRANKLIN, University of Washington, Seattle
GEORGE M. HORNBERGER, University of Virginia, Charlottesville
DEBRA KNOPMAN, Progressive Foundation, Washington, D.C.
PERRY L. McCARTY, Stanford University, California
JUDITH E. McDOWELL, Woods Hole Oceanographic Institution, Massachusetts
S. GEORGE PHILANDER, Princeton University, New Jersey
RAYMOND A. PRICE, Queen's University at Kingston, Ontario
THOMAS C. SCHELLING, University of Maryland, College Park
ELLEN K. SILBERGELD, University of Maryland Medical School, Baltimore
STEVEN M. STANLEY, The Johns Hopkins University, Baltimore, Maryland
VICTORIA J. TSCHINKEL, Landers and Parsons, Tallahassee, Florida

Staff

STEPHEN RATTIEN, Executive Director
STEPHEN D. PARKER, Associate Executive Director
MORGAN GOPNIK, Assistant Executive Director
GREGORY SYMMES, Reports Officer
JAMES MALLORY, Administrative Officer
SANDI FITZPATRICK, Administrative Associate
SUSAN SHERWIN, Project Assistant

The National Academy of Sciences is a private, nonprofit, self-perpetuating society of distinguished scholars engaged in scientific and engineering research, dedicated to the furtherance of science and technology and to their use for the general welfare. Upon the authority of the charter granted to it by the Congress in 1863, the Academy has a mandate that requires it to advise the federal government on scientific and technical matters. Dr. Bruce Alberts is president of the National Academy of Sciences.

The National Academy of Engineering was established in 1964, under the charter of the National Academy of Sciences, as a parallel organization of outstanding engineers. It is autonomous in its administration and in the selection of its members, sharing with the National Academy of Sciences the responsibility for advising the federal government. The National Academy of Engineering also sponsors engineering programs aimed at meeting national needs, encourages education and research, and recognizes the superior achievements of engineers. Dr. William A. Wulf is interim president of the National Academy of Engineering.

The Institute of Medicine was established in 1970 by the National Academy of Sciences to secure the services of eminent members of appropriate professions in the examination of policy matters pertaining to the health of the public. The Institute acts under the responsibility given to the National Academy of Sciences by its congressional charter to be an adviser to the federal government and, upon its own initiative, to identify issues of medical care, research, and education. Dr. Kenneth I. Shine is president of the Institute of Medicine.

The National Research Council was organized by the National Academy of Sciences in 1916 to associate the broad community of science and technology with the Academy's purposes of furthering knowledge and advising the federal government. Functioning in accordance with general policies determined by the Academy, the Council has become the principal operating agency of both the National Academy of Sciences and the National Academy of Engineering in providing services to the government, the public, and the scientific and engineering communities. The Council is administered jointly by both Academies and the Institute of Medicine. Dr. Bruce Alberts and Dr. William A. Wulf are chairman and interim vice-chairman, respectively, of the National Research Council.

Foreword

The discipline of geography has been undergoing a renaissance in the United States during the past decade. Geography's research focus on the study of human society and the environment through the perspectives of place, space, and scale is finding increased relevance in fields ranging from ecology to economics. At the same time, many of its research tools and analytical methods have moved from the research laboratory into the mainstream of science and business. Geography is undergoing a rebirth in education as well—it has become an organizing framework for presenting a wide variety of classroom subjects. It is recognized as an important subject in American schools, and enrollments in geography programs in American colleges and universities are increasing sharply to meet demands from employers for geographically literate students.

Rediscovering Geography: New Relevance for Science and Society is the first comprehensive assessment of geography in the United States in almost 30 years. It provides a broad overview of the discipline and shows how its perspectives and tools are being used by educators, business people, researchers, and policymakers to address a wide range of scientific problems and societal needs. It also provides recommendations for strengthening the discipline's intellectual and institutional foundations to meet growing demands for geography-based knowledge, education, and expertise. The report illustrates that good science and societally relevant science need not be mutually exclusive endeavors, and it enables us to see clearly the societal benefits of scientific research.

The National Research Council very much appreciates the support this study received from the Association of American Geographers, the Environmental

Systems Research Institute, the National Geographic Society, the National Science Foundation, the U.S. Bureau of the Census, the U.S. Department of Transportation, the U.S. Environmental Protection Agency, and the U.S. Geological Survey.

Bruce Alberts
Chairman
National Research Council

Preface

Any academic discipline is a means, not an end. It is a means for such intellectual ends as learning, knowing, and understanding. It is a means for such social ends as progress and problem solving. It is a means for such individual ends as opportunity and fulfillment. Sometimes, we get so caught up in a search for paradigms that perpetuate our disciplinary identities that we forget why it is that we are supported to do the jobs that we do. But we can expect to be reminded in coming years that, just as the U.S. federal government is rethinking functional subdivisions that date back many generations in preparing for a new century with limited resources, the academic world will also be rethinking how it is subdivided and whether new approaches might be better for reaching our collective ends.

In such a time of reflection and change, this report is about geography as a means rather than as an end. It is about subject matter, tools, and perspectives rather than about an academic discipline as such, directed mainly toward readers outside geography whose interest is more in what geography can offer to their concerns than in how geography thinks of itself.

The reason for carrying out this assessment at this particular time is a well-documented growing perception (external to geography as a discipline) that geography is useful, perhaps even necessary, in meeting certain societal needs. As a result, many parties concerned about the ends of academic science have been asking more from the information, techniques, and perspectives associated with geography than the nation's scientific and educational systems are delivering; and the gap between demand and supply may be widening. The most salient aspect of this demand is in education reform, especially in grades K–12, where geography education is indeed expanding rapidly. An expansion in classroom instruction without a strong foundation in knowledge and skills, however, does not serve the ends of education well; and it is important to balance this one kind of response to external demands with an assessment of the knowledge base that undergirds it.

More generally, in fact, it makes sense to start seeking answers to the geography demand-supply gap by getting better informed about what geography as a scientific discipline is and does, both to understand the place of the disciplinary infrastructure in meeting broader needs and to understand some of the dimensions of the still inchoate questions arising externally. Furthermore, in struggling with relationships between transdisciplinary ends and disciplinary means, in this case where geography is concerned, the broader scientific and societal communities can advance the process of reconsidering the relevance of conventional academic disciplines as we all look toward the future.

With these objectives in mind, the Rediscovering Geography Committee was established in 1993 to perform a comprehensive assessment of the discipline of geography in the United States. The assessment is intended to convey to a wide range of readers, especially scientists and decision makers who are not geographers, the substance of geography as a subject—which is generally not well understood beyond a relatively small group of professional practitioners—and to identify ways to make the discipline more relevant to science, education, and decision making.

In performing its assessment the committee held five meetings during a period of about 12 months to gather information, debate the issues, and develop this report. The committee made a special effort to communicate with the geography community during the study and to solicit the advice of individual geographers. To this end, the committee published notices in the monthly newsletter of the Association of American Geographers (AAG), held special sessions at the 1994 and 1995 AAG annual meetings, and met with several groups of geographers who indicated concerns about the committee's deliberations. Although many geographers provided valuable input to our assessment, this report is a consensus document of the committee, and the committee is solely responsible for its content, conclusions, and recommendations.

On behalf of the committee, I would like to recognize the efforts of many individuals and organizations who contributed to the successful completion of this report. First of all, the committee would never have come into existence without the leadership of the U.S. National Committee for the International Geographical Union, and especially Melvin Marcus, its chair through 1992, who first identified the need for this reassessment given the mounting scientific and societal demands on geography as a discipline. Mel, Bill Turner, and Tony de Souza contributed to developing the project concept, and Reds Wolman, Julian

Wolpert, and Ron Abler were instrumental in refining it and obtaining approval for the study to be carried out under the auspices of the National Research Council.

The committee's effort would not have been possible without generous financial support from a variety of agencies and organizations interested in both geography's ends and means. We gratefully acknowledge support from the Association of American Geographers, the Environmental Systems Research Institute, the National Geographic Society, an initial grant for planning purposes from the Governing Board of the National Research Council, the Anthropological and Geographic Sciences Program of the National Science Foundation, the Decennial Census Program of the U.S. Bureau of the Census, the Bureau of Transportation Statistics of the U.S. Department of Transportation, the Environmental Monitoring and Assessment Program of the U.S. Environmental Protection Agency, and the National Mapping Division of the U.S. Geological Survey.

The committee also expresses its gratitude to the many geographers who provided advice and materials for the report, too numerous to list, to several colleagues who provided informal reviews of early drafts of selected chapters (Ron Abler, Brian Berry, Michael Dear, Rodney Erickson, Susan Hanson, and Joel Morrison), and to the anonymous reviewers enlisted by the National Research Council. We were assisted by the AAG Employment Forecasting Committee—Pat Gober (Chair), Amy Glasmeier, James Goodman, David Plane, Howard Stafford, and Joseph Wood—who, at the request of the committee and with the support of the AAG, produced materials on the employment trends in geography that appear in Appendix A. We are grateful to the AAG for its permission to include copies of journal articles by that committee in the appendix.

Finally, the committee and I thank the staff of the Board on Earth Sciences and Resources of the National Research Council for its help in obtaining financial support for the study, facilitating the meetings, and working closely with the committee to produce this report. In the early stages we were ably assisted by Bruce Hanshaw and Shelley Myers. In the latter stages of committee meetings and throughout the period of report preparation, we benefited from the superb administrative assistance of Jenny Estep, who was a model of professionalism, productivity, and grace under pressure. Most profoundly, we are indebted to Kevin Crowley, the study director. Kevin went far beyond the call of duty in contributing to the report intellectually as well as administratively, in improving its communication to audiences external to geography, and in caring about the quality of both the process and the product. We on the committee consider him a colleague and a friend in the fullest sense—and an honorary geographer who sometimes helped us to see things in our own messages that we ourselves had missed. His conscientious, diligent, and thoughtful participation reflects great credit on the National Research Council.

Thanks to you all, and my personal thanks to a fine committee who very quickly overcame differences in subdisciplinary acculturation and personal self-interest to join together so effectively in a common enterprise. Every individual member made a significant contribution, and I believe that all of us were enriched by the opportunity to work together with such distinguished colleagues and interesting people. None of us is entirely happy with every detail that emerged from the consensus process, but the overall orientation of the report represents a unanimous judgment, and the report itself speaks for a strong consensus among a disparate group of strong-minded individualists, all of them experts in some part of geography's intellectual territory. We trust that our assessment will serve as a stimulus for a continuing discussion of the ends and means of geography, together with strategies for strengthening their connections, because we certainly do not consider it the last word.

Thomas J. Wilbanks, *Chair*
Rediscovering Geography Committee

Contents

Executive Summary

In the past decade, concerns about "geographic illiteracy" have been the catalyst for a new focus on geography in the United States. Recent calls to "do something" about geographic illiteracy in this country can be traced to concerns about U.S. competitiveness in the global economy, combined with surveys that documented an astonishing degree of ignorance in the United States about the rest of the world. There is a growing public recognition that our national well-being is related to global markets and international political developments, to the continued prominence of environmental issues in social discourse, and to the emergence of computer and telecommunications technologies that emphasize graphic images such as maps and other spatial diagrams—all of which are associated in the public's mind with geography.

One result of this increased attention is a rediscovery of the importance of geography education in the United States. Geography is identified as a core subject for American schools, on a par with science and mathematics, in a series of recent policy statements and legislative proposals for national education reform. These include the report of the Charlottesville, Virginia, Summit convened by the 50 state governors and President Bush in October 1989; education reform plans of both the Bush and the Clinton administrations; and *Goals 2000: The Educate America Act*, passed by Congress in March 1994.

Geography has also been rediscovered by students. In the period 1986/1987 to 1993/1994, the number of undergraduate majors in geography grew by an estimated 47 percent nationwide and by 60 percent in Ph.D.-granting departments. Between 1985 and 1991, graduate program enrollments in geography grew by

33.4 percent, compared with a 15.3 percent increase in the social sciences and a 5.4 percent decrease in the environmental sciences.

This process of rediscovery has been mirrored in the research community as well. Research at the frontiers of fields as diverse as planning, economics, finance, social theory, epidemiology, anthropology, ecology, environmental history, conservation biology, and international relations has highlighted the importance of geographic perspectives. In particular, the importance of spatial perspectives—through such notions as *place* and *scale*—is being recognized in many fields, extending the influence of geography well beyond its relatively small group of professional practitioners.

The increased use of perspectives, knowledge, and techniques associated with a relatively small academic discipline raises several questions for the scientific community. Most directly, what is geography, and how does it connect with broad concerns of society and science? If geography is to play a more prominent role in education and decision making, do its scientific foundations need to be strengthened in order to support its expanded responsibilities?

With these questions in mind, the National Research Council established the Rediscovering Geography Committee to perform a comprehensive assessment of geography in the United States. The objectives of this assessment are:

1. to identify critical issues and constraints for the discipline of geography,
2. to clarify priorities for teaching and research,
3. to link developments in geography as a science with national needs for geography education,
4. to increase the appreciation of geography within the scientific community, and
5. to communicate with the international scientific community about future directions of the discipline in the United States.

In addressing these issues, this report focuses on broad national and global themes in science and society, geography's potential as a perspective and a body of knowledge to help address these themes, and constraints on geography's capability as an academic discipline to respond. As examples, it draws mainly on experience from within geography as a discipline, although valuable geographic work is done outside the discipline as well, because the committee was comprised very largely of professional geographers. Where possible, however, the examples are selected to illustrate the interconnectedness between disciplines that characterizes so much geographic investigation and facilitates the flow of ideas, concepts, and techniques across disciplinary boundaries.

THE PERSPECTIVES, SUBJECT MATTER, AND TECHNIQUES OF GEOGRAPHY

To most Americans, geography is about place names. Concerns about geographic ignorance usually focus on people's inability to locate cities, countries, and rivers on a world map, and geographic instruction is often equated with conveying information about remote parts of the world. From this perspective it may be a surprise to some that the discipline of geography has a great deal to say about many of the critical issues facing society in the late twentieth century.

Geographers are engaged in valuable research and teaching on matters ranging from environmental change to social conflict (see Chapter 2). The value of these activities derives from the discipline's focus on the evolving character and organization of the Earth's surface; on the ways in which interactions of physical and human phenomena in space combine to create regions with distinctive natural and (or) social characteristics, or *places*; and on the influences those places have on a wide range of natural and human events and processes. Such concerns are not simply exercises in expanding the encyclopedic knowledge of faraway places; they go to the heart of some of the most urgent questions before decision makers today.

A central tenet of geography is that "location matters" for understanding a wide variety of processes and phenomena. Indeed, geography's focus on location provides a cross-cutting way of looking at processes and phenomena that other disciplines tend to treat in isolation. Geographers focus on "real-world" relationships and dependencies among the phenomena and processes that give character to a place. Geographers also seek to understand relationships among places: for example, the flows of peoples, goods, and ideas that reinforce differentiation or enhance similarities. In other words, geographers study both the "vertical" integration of characteristics that define place and the "horizontal" connections between places. Geographers also focus on the importance of scale (in both space and time) in these relationships. The study of these relationships has enabled geographers to pay attention to complexities of places and processes that are frequently treated in the abstract, if at all, by other disciplines.

Geography's perspectives are supported by a body of distinctive techniques for *observation,* such as field exploration, remote sensing, and spatial sampling, and for the *analysis and display* of geographic data, such as cartography, visualization, spatial statistics, and geographic information systems (GISs; see Chapter 4). These techniques are shared with other disciplines, but geography has contributed fundamentally to their development and improved application.

The traditional tool in geography for the display of spatially referenced information is the map. To many, the term "map" connotes a fixed, two-dimensional paper product containing point, line, and areal data. During the past generation, however, advances in data collection, storage, analysis, and display have made this traditional view obsolete. The modern map is a dynamic and multidimensional product that exists in digital form, opening up new areas of research and application for geographic investigation. This research has led to the development of GISs, which, along with techniques for geographic visualization and methods of spatial analysis, facilitate an increasingly complex and contextual understanding of the world. Current research in GISs is expanding the technique to incorporate more advanced geographic concepts and analysis methods.

GEOGRAPHY'S CONTRIBUTIONS TO SCIENTIFIC UNDERSTANDING AND DECISION MAKING

Geography offers significant insights into some of the major questions facing both the pure and applied sciences. In addition, as society itself is recognizing, many of the major questions facing society at the local, national, and international scales have very important geographic dimensions.

Geography's traditional interest in integrating phenomena and processes in particular places, for example, has a new relevance in science today, in connection with the search for what some have called a "science of complexity." In its explorations as a science of flows, geography has been a leader in understanding spatial interactions, a subject of broad interest to both science and society. Moreover, geography's long-standing concern with interdependencies among scales is relevant to discussions across the body of science of relationships between microscale (small or local) and macroscale (large or global) phenomena and processes (see Chapter 5).

Geographic perspectives and techniques have found important applications in decision making in both the private and the public sectors, especially as global economic and environmental issues and modern information technologies have grown in importance. Geographers have made significant contributions to decision making at local, regional, and global scales for a wide variety of issues—for example, management of hazards, understanding global environmental and economic changes and their interactions with local changes, and developing effective business strategies (see Chapter 6).

STRENGTHENING GEOGRAPHY'S FOUNDATIONS

The ability of geographers to respond to the growing demand for its skills and perspectives is limited by several realities (see Chapter 7). Despite three decades of growth in the number of professional geographers, the geography community remains small relative to most other natural and social science disciplines. Few colleges and universities have large geography departments, and many institutions of higher learning have no geography programs at all, including some of the nation's leading universities. This situation is extraordinary by world standards because geography is a core subject in most universities in Europe and East Asia. Additionally, women and minorities are underrepresented in senior academic and professional positions relative to their numbers in the general population, and, at present, few minorities are entering the field. This small human and programmatic base will make it difficult for the discipline to respond effectively to increased demands for attention—demands that are likely to increase still further in the years ahead, especially in education.

Realizing geography's potential requires more than addressing the problems presented by the discipline's small size and limited diversity, however. In several critical subject areas, geography's intellectual foundations need to be strengthened to ensure that its contributions to science and society are solidly grounded. The discipline needs to strengthen its understanding of complex systems;[1] interactions between scales; interactions between society and nature; and geographic learning, including the effectiveness of interactive learning tools on geographic education. At least as important, the appreciation and use of geography by nongeographers need to be fostered, so that the capacity to make use of the discipline's perspectives, knowledge, and techniques grows along with the capacity of the discipline to supply them. This includes enhancing the geographic competency of the general population and fostering better geographic training in colleges and universities.

Filling these gaps will require external support of types and at levels beyond those that have been characteristic in the past, in a setting where conventional sources of support will be constrained by external circumstances. Looking toward the next century, it seems clear that realizing geography's potentials will require innovative new partnerships between provider and user, supported and supporter, one science and another, and basic research and applications of knowledge.

If geography as a discipline can be a pathfinder in developing and fulfilling such partnerships, it can play a significant role in realizing its potential, without depending entirely on external action. But in doing so the discipline faces its own internal challenges. In order to respond to external demands and to gain additional external support, the discipline needs to place increased emphasis on such traditional strengths as integration in place, field observation, and foreign field research, as well as geography education as a challenge for research and practice. It also needs to promote more professional interactions with other scientific disciplines and with users of geographic knowledge in government and business at all levels. And it needs to enhance not only its diversity as a discipline but also its appreciation for diversity.

The Rediscovering Geography Committee has concluded that a number of internal and external actions are needed to strengthen the discipline and thereby increase its contributions to science and society in the United States in the coming decades. Chapter 8 lists the full set of conclusions. The committee's 11 recommendations are divided into three categories oriented toward the external audiences of this report, including one recommendation about the process of implementing the previous 10:

---

[1] The term *complex* is used to describe processes or systems that exhibit nonlinear (i.e., multiplicative or exponential) or chaotic (i.e., unpredictable) behaviors. Many processes and systems studied by geographers (e.g., climate, stream-network, ecosystem, and landscape systems) exhibit complex behaviors.

To improve geographic understanding:

1. Increased research attention should be given to certain core methodological and conceptual issues in geography that are especially relevant to society's concerns.

2. More emphasis should be placed on priority-driven, cross-cutting projects.

3. Increased emphasis should be given to research that improves the understanding of geographic literacy, learning, and problem solving and the roles of geographic information in education and decision making, including interactive learning strategies and spatial decision support systems.

To improve geographic literacy:

4. Geography education standards and other guidelines for improved geography education in the schools should be examined to identify subjects where geography's current knowledge base needs strengthening.

5. A significant national program should be established to improve the geographic competence of the U.S. general population as well as of leaders in business, government, and nongovernmental interest groups at all levels.

6. Linkages should be strengthened between academic geography and users of its research.

To strengthen geographic institutions:

7. A high priority should be placed on increasing professional interactions between geographers and colleagues in other sciences.

8. A specific effort should be made to identify and address disparities between the growing demands on geography as a subject and the current capabilities of geography to respond as a scientific discipline.

9. A specific effort should be made to identify and examine needs and opportunities for professional geography to focus its research and teaching on certain specific problems or niches, given limitations on the human and financial resources of the discipline.

10. University and college administrators should alter reward structures for academic geographers to encourage, recognize, and reinforce certain categories of professional activity that are sometimes underrated.

To encourage implementation of these recommendations:

11. Geographic and related organizations—especially the Association of American Geographers, National Geographic Society, National Science Foundation, and the National Research Council—should work together to develop and execute a plan to implement the recommendations in this report.

1

Introduction

In the past decade, concerns about "geographic illiteracy" have been the catalyst for a new focus on geography in this country. Our future as a nation depends substantially on our knowledge base, and many observers agree that current problems with productivity and competitiveness can be traced in large part to deficiencies in this knowledge base among our fellow citizens. One of the most glaring of these deficiencies is in our knowledge of geography, which is the reason for this report.

Recent calls to do something about geographic illiteracy in the United States can be traced to concerns in the 1980s about U.S. competitiveness in the global economy, combined with surveys that documented an astonishing degree of ignorance in the United States about the rest of the world. For example, in a 1986 survey of adults in nine countries, young U.S. adults knew the least about geography of any age group in any country. About one-half could not point out South Africa on a map or identify even one South American country, and only 55 percent could locate New York (Gallup Organization, Inc., 1988). Similarly, a 1987 survey of 5,000 high school seniors in seven cities found that one-quarter of Dallas students could not name the country bordering the United States on the south (Gallup Organization, Inc., 1988).

Since the mid-1980s, calls for attention to geographic illiteracy have been frequent, not only from academia and the federal government but from business and state government as well. Consider the following examples:

> We as a nation are constantly surprised by world political and economic events. They occur in places we never heard of for reasons we do not understand. And

> we often do not realize the importance of these events in our daily lives. . . . We must accept the fact that we are as dependent on other nations as they are on us, and we must begin to understand our global neighbors. . . . The problem is that we often do not teach geography in this country, and when we do, it is frequently taught poorly.[1] (Southern Governors' Association, Cornerstone of Competition, November 1986)

> I was disturbed by a new survey that shows most Americans don't know where to find the major trouble spots of the world. . . . Before we can figure out how to stop people from stealing our jobs or sending us their illegal drugs, we at least had better find out where they are. (Clarence Page, *Chicago Tribune*, July 31, 1988)

> The United States is not well-prepared for international trade. . . . How are we to open overseas markets when other cultures are only dimly understood? The imperatives are clear: It is time to learn languages. It is time to learn geography. It is time to change our thinking about the world around us. For we cannot compete in a world that is a mystery beyond our borders. (National Governors' Association, *America in Transition: The International Frontier*, 1989)

> Geographic information is critical to promote economic development, improve our stewardship of natural resources, and protect the environment. (Presidential Executive Order, Coordinating Geographic Data Acquisition and Access: The National Spatial Data Infrastructure, April 11, 1994)

Behind these calls for increased attention to geographic illiteracy in a very broad sense is a growing public recognition that our national well-being is related to global markets and international political developments, the continued prominence of environmental issues in social discourse, and the emergence of computer and telecommunications technologies that emphasize graphic images such as maps and other spatial diagrams.

One result of this increased attention is a rediscovery of the importance of geography education in the United States. Geography is identified as a core subject for American schools, on a par with science and mathematics, in a series of recent policy statements and legislative proposals for national education reform. These include the report of the Charlottesville summit convened by the 50 state governors and President Bush in October 1989; education reform plans of both the Bush and the Clinton administrations; and *Goals 2000: The Educate America Act*,[2] passed by Congress in March 1994.

Geography has also been rediscovered by students. In the period 1986/1987 to 1993/1994, the number of undergraduate majors in geography grew by an estimated 47 percent nationwide and by 60 percent in Ph.D.-granting departments.

---

[1]In 1987 only 15 percent of high school graduates had completed a course in world geography (National Center for Education Statistics, 1993), and other sources indicate that many of the courses were taught by instructors with little or no training in geography.

[2]P.L. 103/227, March 31, 1994.

Between 1985 and 1991, geography graduate program enrollments grew by 33.4 percent, compared with a 15.3 percent increase for the social sciences and a 5.4 percent decrease for the environmental sciences (see Figure 1.1 and Appendix A).

This process of rediscovery has been mirrored in the research community as well. Research at the frontiers of fields as diverse as planning, economics, finance, social theory, epidemiology, anthropology, ecology, environmental history, conservation biology, and international relations has been highlighting the importance of geographic perspectives (e.g., Giddens, 1984; Cliff and others, 1986; Forman and Godron, 1986; Krugman, 1991; Soule, 1991; Ruggie, 1993). The importance of a geographic perspective—through recognizing the critical importance of such notions as place and scale—is being acknowledged in many fields, extending the influence of geography well beyond its relatively small group of professional practitioners.

This increased emphasis on the perspectives, knowledge, and tools associated with a relatively small academic discipline raises several questions for the scientific community. Most directly, what is geography, and how does it connect with the broad concerns of society and science? Also, if geography is to play a more prominent role in education and decision making, do its scientific foundations need to be strengthened in order to support its expanded responsibilities?

With these questions in mind, in 1993 the National Research Council (NRC) established the Rediscovering Geography Committee to perform a comprehensive

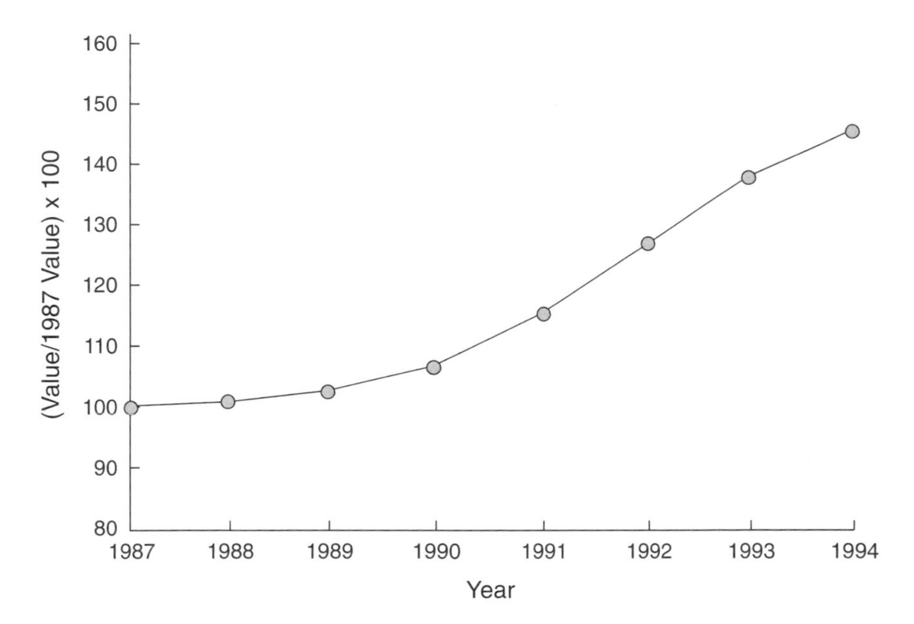

FIGURE 1.1 Undergraduate enrollments in geography, 1987/1994. Source: Appendix A, this report.

assessment of geography in the United States. The objectives of the assessment are the following:

1. to identify critical issues and constraints for the discipline of geography,
2. to clarify priorities for teaching and research,
3. to link developments in geography as a science with national needs for geography education,
4. to increase the appreciation of geography within the scientific community, and
5. to communicate with the international scientific community about the future directions of the discipline in the United States.

CONTEXT OF THE REPORT

This assessment was conducted during a time of widespread change in conditions both external and internal to the discipline. As a result, *change* became a central theme of the committee's deliberations. The deliberations resulted in this report, which differs significantly from previous NRC assessments of the discipline.[3] Earlier assessments focused internally on disciplinary paradigms and vocabulary. This report instead is focused outward on broad national and global issues; geography's potential as a body of knowledge, perspectives, and techniques to help address them; and constraints on geography's capability as an academic discipline to respond. It is written to articulate to the scientific and policy-making communities geography's relevance to such issues, to assist the discipline itself in strengthening its connections with them, and to spotlight the roles of scientific knowledge and skills in geography's response to external expectations.

In order to put the report in perspective, the changes considered by the committee are worth reviewing.

Changes in Society

During the past decade, American society has been profoundly affected by global geopolitical, political-economic, and environmental changes. In this dramatic period, political and economic reforms in the former Soviet Union and Central Europe ended the Cold War, which dominated international relations for nearly half a century. The Pacific Rim and Western Europe have become potent competitors for international and U.S. domestic markets, creating new concerns about the U.S. trade balance and U.S. jobs. Market reform and democratization in many areas—Eastern Europe, India, China, Mexico, South Africa, and else-

[3]Compare, for example, with two NRC assessments of geography in the 1960s: *The Science of Geography* (NRC, 1965) and *Geography* (Taaffe et al., 1970).

where—have changed international political-economic relationships. Scientific evidence of a thinning of the Earth's ozone layer has led to a new sensitivity to trends in global environmental change, and further evidence of accelerated changes in our physical environment—for example, land and water pollution, deforestation, and desertification—has triggered a general concern about "sustainable development." In fact, uneasiness about environmental changes, local and distant, is having an unprecedented impact on policy agendas and market conditions worldwide. In addition, technological change has produced a revolution in information delivery and communication, as powerfully demonstrated during the Gulf War of 1991. Few periods in world history have seen such widespread fundamental change. Although these changes have focused welcome attention on geography as a subject, geography as an academic discipline is limited by size and other constraints from contributing its knowledge, perspectives, and techniques to improving the nation's ability to cope with and prosper under these changing conditions.

Changes in Relationships Between Society and Science

American society has grown increasingly skeptical about the wisdom and value of science, as traditionally defined. One reason may be that advances in science have not been matched by advances in the human condition (e.g., Handler, 1979). Another may be that society expects science to reduce uncertainties, when in so many cases it has instead increased uncertainties. At any rate, science is now being held accountable for its payoffs, during a period when public funds to support research and education are increasingly scarce (NRC, 1993a). The era when public support for science could be expected to increase more rapidly than the nation's rate of economic growth is over, at least for now (Gibbons, 1994), and science is being measured against its usefulness in improving the human condition (OSTP, 1994). Although geography is not accustomed as an academic discipline to thinking in these terms, it has a history of relatively close links between basic research and societal issues. This experience can be useful to the scientific community, as well as to geography itself, under the new conditions for support of science.

Changes in Relationships Between Society and Geography

Perhaps the most dramatic indication of changes in geography's external environment has been the emergence of a strong grass-roots demand for geography education for the first time in U.S. history. Without reviewing the history of *Goals 2000: The Educate America Act* in any detail, it is clear that geography is being asked to meet educational needs at kindergarten through grade 12 (K–12) levels that extend beyond geography as an academic discipline per se. In many respects, geography is being seen as an umbrella under which students are taught broadly about interconnections in the contemporary world. Although this spotlight is most welcome from the standpoint of a discipline that for decades felt that it received too little attention, it comes at a time when most universities and institutions that support research face severe financial stringencies, limiting their ability to provide the resources needed to meet the increased expectations from geography that are equivalent to those of much larger bodies of science.

Changes Within Geography Itself

Finally, since the previous NRC assessments, geography in the United States has become larger and more prominent. For example, since 1960 the membership of the Association of American Geographers (AAG) has grown from 2,000 to more than 7,000, and the number of geographers elected to the National Academy of Sciences has increased from zero to eight. Geography has changed in its central thrusts as a discipline, moving toward emphases articulated by Robert Kates as president of the AAG in 1993/1994: improving geographic literacy, relating geographic scholarship to social needs, and strengthening connections with others (Kates, 1994a). The discipline has become more issue oriented in its research agendas, and it has directed more of its attention to moral dimensions of research questions.[4] Such major geographic organizations as the AAG and the National Geographic Society (NGS) have moved toward closer associations, and all of geography's national associations (AAG, NGS, the National Council for Geographic Education, and the American Geographical Society) have come together to promote initiatives in geographic education through the Geography Education National Implementation Project.[5]

Many of these changes within geography are themselves responses to changes in society, and some of them have affected the ways professional geographers view the search for knowledge. Although this report is about geography as a science, such a focus is itself different from what it was a generation ago (see Sidebar 1.1).

At the same time, geography (like other disciplines) has been shaped by its access to resources for research and teaching. For instance, the focus in the late 1960s and early 1970s on U.S. social and environmental problems, combined with a steep reduction in financial support for foreign-area research, reduced the proportion of younger American geographers pursuing field research in other countries. In addition, the rapidly growing importance of technologies for information gathering, analysis, and display has increased the costs of staying at the frontier in many fields of geographic research.

Taken together, these changes are both so profound and so recent that,

---

[4]In the sense of research that provides scientifically valid methods to investigate many of the moral questions of concern to society.

[5]For additional information about geography's organizations, see Appendix B.

SIDEBAR 1.1 Geography's Approaches to Learning

Caught up in a world of change, geography has been extending and diversifying its ways of seeking knowledge, that is, its *epistemologies.* The ascendance of traditional scientific methods in the 1960s brought a new emphasis on theory within both the subject and the discipline, and these traditions have matured and progressed since then, especially in physical geography. Meanwhile, a full complement of other approaches being explored in the social sciences and humanities have also found expression in geography, partly because geography's subject matter is so wide ranging, and partly because of concerns across the research world about claims of neutrality or objectivity associated with any particular path toward knowledge. For example, "social theory" has had a major impact on geographic research, emphasizing the societal context of historical processes, and geographers have struggled with arguments that all "scientific" theory and observations are socially constructed and that all interpretations are contingent on the social context of the analyst (see the last section in Chapter 3, "Geographic Epistemologies"). Professional geographers also carry out research that is not intended to be scientific but is anchored in the humanities. Insights from such research are often valuable sources of ideas for geography as a science, and they remind geographers of the power of imagination and narrative in pursuing understanding.

This report takes an eclectic and inclusive view of geography as a science, emphasizing the relevance of research results more than the procedures used to derive them. It reflects the committee's view, permeating many of the sciences today, that multiple paths toward understanding are worth exploring and that breakthroughs in learning are likely to be fostered by a dialogue among many different paths.

looking toward a new century, the committee believes that its task calls for breaking new ground.

SCOPE OF THE REPORT

A further issue for the committee was its interpretation of relevance, in terms of external expectations on the academic discipline of geography. Rather than limiting the scope of its work to geographic illiteracy, narrowly defined (e.g., the role of geography in disseminating basic facts about foreign areas), the committee examined geography's current and potential connections with a broader range of societal and scientific challenges and opportunities as the twentieth century draws to a close.

As one example of such a broader agenda, the National Science Foundation recently identified eight strategic fields of research, education, and information transfer, associated with U.S. national objectives identified by the President's National Science and Technology Council (NSF, 1994). Of these, five are fields in which geography should be a central contributor: global change research;

environmental research; high-performance computing and communications (e.g., geographic information systems and visualization); civil (public) infrastructure systems; and science, mathematics, engineering, and technology education consonant with the Educate America Act. Geography is also relevant in more subtle ways to the other three fields—biotechnology, advanced materials and processes, and advanced manufacturing technology—through its focus on environmental and social issues, resource use, locational decisions, and technology transfer.

At the same time, as it seeks to improve our understanding of these issues and the more basic questions that underlie them, science is confronting certain fundamental issues across a wide range of disciplines that seem conceptually similar. As a part of science, geography is deeply involved in some of these issues, such as relationships between macroscale and microscale phenomena and processes,[6] understanding complex systems,[7] developing integrative approaches to understanding complexity, and understanding relationships between form and function.

This report examines geography's current and potential relevance to these kinds of issues for science as well as to salient issues for society.

CONTENT OF THE REPORT

To address these questions, Chapter 2 of this report offers several brief examples of geography's relevance to critical issues for U.S. and international society, laying a foundation for later chapters. Chapter 3 summarizes the perspectives of geography as it addresses these and other issues, and Chapter 4 describes geography's techniques. Chapters 5 and 6 then turn in somewhat more detail to geography's potential to contribute first to scientific understanding related to critical issues and then to decision making related to such issues. Chapter 7 confronts certain needs for research and learning initiatives in order to strengthen the discipline's foundations if it is to respond effectively to the changes that confront it, including the unprecedented demands to support educational reform in the United States. Finally, Chapter 8 presents the committee's conclusions and recommendations related to research, education, and outreach. Appendix A reports available data on education and employment trends in geography.

As noted previously, this report is written partly to address the interests

[6] *Macroscale* and *microscale* define the end points of a continuum of spatial geographic scales often referred to as the *local-global continuum*. This continuum ranges from very local (e.g., a town or a city block) to global (i.e., at the scale of the Earth). Geographers generally divide this continuum into three segments: micro-, meso-, and macroscale. These are roughly equivalent to local, regional, and global scales as defined by conventional usage.

[7] The term *complex* is used to describe processes or systems that exhibit nonlinear (i.e., multiplicative or exponential) or chaotic (i.e., unpredictable) behaviors. Many processes and systems studied by geographers (e.g., climate, stream-network, ecosystem, and landscape systems) exhibit complex behaviors.

and concerns of nongeographers about geography's subject matter rather than geography as a discipline (see Sidebar 1.2). It does not review the current state of geography to inform the discipline itself. It is not a description of the history of the discipline in the United States or of how its history has been different in other countries. It is not the statement of a disciplinary consensus on the issues that are addressed. It does not provide a comprehensive review of geography's literature. Indeed, the committee made a conscious effort to keep referencing to a minimum.[8]

Instead, this report is written for the broad audience that is curious about geography's new place in a national spotlight. It reflects the consensus of the committee on how geography can contribute to issues for science and society on the threshold of the twenty-first century.

SIDEBAR 1.2 Geography, Geographer, Geographic: What's in a Name?

The knowledge, perspectives, and techniques of geography (i.e., geography's *subject matter*), which are discussed in Chapters 3 and 4 of this report, have found wide application in many scientific fields, and they are practiced by more than professional geographers alone. In preparing this assessment of the relevance of geography to science and society, the committee focused on this *subject matter* because that is the essence of the rising external interest, rather than on the discipline per se.

The examples used in this report were largely taken from the geography literature with which the committee is most familiar, although some references are also made to work by scientists who would not call themselves geographers nor refer to their work as "geographic," even though it concerns geography's subject matter. The committee has made no effort to define the boundaries of geography as an academic discipline because the boundaries are diffuse and unlikely to be of much interest to the nongeography audience for this report. Nor has the committee sought to lay claim to an expanded boundary for geography as a discipline by claiming the geographic work of other disciplines as its own. Rather, the committee's intention is to illustrate the application of geography's subject matter wherever it is done—and thereby to demonstrate geography's interconnectedness to other scientific disciplines, which hastens the flow of ideas, concepts, and techniques, and to encourage a stronger influence of geography far beyond its small group of academic practitioners.

In this report the committee uses the term *geography* to refer to the academic discipline and its subject matter, some of which is shared with other natural and social science disciplines. The term *geographer* is used to refer to practitioners of geography who have acquired expertise in the discipline's knowledge, perspectives, and techniques, either through academic training or other professional experience. The term *geographic* is used to differentiate the subject matter of geography from the academic discipline.

---

[8]For more information about the discipline's literature, see Abler et al. (1992) and Gaile and Willmott (1989).

2

Geography and Critical Issues

To most Americans, geography is about place names. Concerns about geographic ignorance usually focus on people's inability to locate cities, countries, and rivers on a world map, and geographic instruction is often equated with conveying information about remote parts of the world. From this perspective it may be a surprise to some that geography has relevance to many of the critical issues facing society in the late twentieth century.

Geographers and others using geographic knowledge and perspectives, in fact, are engaged in valuable research and teaching on matters ranging from environmental change to social conflict. The value of these activities derives from geography's focus on the evolving character and organization of the Earth's surface, on the ways in which the interactions of physical and human phenomena in space create distinctive places and regions, and on the influences those places and regions have on a wide range of natural and human events and processes. Such concerns are not simply exercises in expanding the encyclopedic knowledge of faraway places; they go to the heart of some of the most urgent questions before decision makers today: How should societies respond to the accelerated pace of environmental degradation in many parts of the world? What are the underlying causes and consequences of the growing disparities between rich and poor? What are the mechanisms that drive the global climate system? What causes the severe floods that have occurred in recent years, and how can society cope with such events? How is technology changing economic and social systems?

Addressing such questions goes far beyond the abilities and insights of any one discipline. Yet each question embodies fundamental geographic dimensions—dimensions that are ignored at society's risk. The geographic perspective is concerned with the significance of place and space on processes and phenomena (see Chapter 3 for a fuller discussion). The geographic perspective motivates such questions as: Why is a particular phenomenon found in some places but not others? What does the spatial distribution of vegetation or homeless people or language traits tell us about how physical and human processes work? How do phenomena found in the same place influence one another, and how do phenomena found at different places influence one another? How do processes that operate at one geographic scale affect processes at other scales? What is the importance of location for efforts to effect (or avoid) political, social, economic, or environmental change?

The importance of the geographic perspective to many contemporary "critical issues" for society is illustrated by a few selected examples in the following sections.

ECONOMIC HEALTH

Perhaps the main reason for American society's strong interest in geography in the 1990s is a sense that jobs, income, and entrepreneurial opportunities in the United States are connected with the global marketplace. The United States is caught up in the profoundly important process of global economic restructuring, in which every nation seeks competitive advantage in providing products and services that global consumers want. U.S. citizens no longer have the highest average standard of living in the world, and many citizens believe that other countries are doing a better job than the United States in responding to new economic conditions. Moreover, U.S. cities and regions are dealing with other dimensions of global economic change, such as reduced military spending with the end of the Cold War and increased interest in environmental sustainability.

Geography is expected to ensure a flow of accurate, timely, and useful information about the rest of the world, but it is more than a repository of place facts. It asks, for example: How and why do commodities, money, information, and power flow from one place to another? What characteristics of a place cause it to do better economically than another? What actions are best taken at national, regional, or local scales to improve economic development? How does global economic change relate to global environmental change?

Geographers contribute to understanding and responding to global economic change through their focus on place and space—in this context, the effects of place (location) and space (the connections between locations at different scales) on economic change and development. For example, Glasmeier and Howland (1995) used the heterogeneous and rapidly growing service sector to study the impacts of advanced information technologies on the growth of rural areas in the United States, recognizing the distinctiveness of rural areas as well as the social, economic, and geographic differences among rural areas. Geographers view nations not only as pieces of a mosaic but as mosaics themselves, that is, of geographically varying combinations of local knowledge and resources. Geographers go beyond regional estimates of production costs and product markets to understand the complex relationships among regional political, social, and environmental conditions and processes. Markusen (1987), for instance, has reviewed the economic and political history of regions and regionalism within the United States to relate political movements and economic structure in an historical and geographic context. Geographers examine location as a factor influencing the connections of particular places to global changes and flows.

A good example of a geographic perspective in action is an analysis of relationships between regional economic growth in the United States and patterns of military expenditures, which was led by Ann Markusen and Peter Hall (Markusen et al., 1991). This analysis suggested that publicly financed industrial production has a different geographic pattern than privately funded industrial activity because of strategic considerations such as the decentralization of production and the importance of relationships among defense contractors, military offices, and congressional budget decisions. Further, it suggested that different periods of military spending have different geographies, but spending in each period has considerable spatial concentration. For instance, "hot wars" such as World War II, Korea, and Vietnam tended to reinforce the prominence of existing industrial centers in the Northeast and Midwest, whereas Cold War spending patterns tended to shift military procurement toward the South, West, and New England. These concentrations make it difficult for regions dependent on military spending to adjust when the nation moves from one period to another.

Such research findings have helped the federal government appreciate the importance of formulating programs to help defense-dependent communities adjust to plant and facility closings and other impacts of defense spending cuts. For instance, these findings have been influential in stimulating initiatives to educate state and local economic development officials about reemployment strategies and options for plant and facility reuse.

ENVIRONMENTAL DEGRADATION

As the twentieth century draws to a close, there is growing concern that humans are irreparably degrading the physical environment that supports them. A wide range of human activities contributes to this problem, including the pollution of air, land, and water as a result of industrial and agricultural activities. In many parts of the world the quality of the air has declined to the point that plant and animal communities are threatened, as well as human health. The heavy use of fertilizers and pesticides in agriculture and the expanding quantity of waste that must be stored on or near the Earth's surface are impairing the quality of the land surface of the planet.

Understanding and confronting the environmental degradation problem requires more than a physical analysis of particular pollutants or an institutional analysis of decision-making structures. It also requires geographic analysis. Why do polluting industries concentrate in particular locations? Where do pollutants go once they leave a factory or dump? Where are the best places to locate polluting industries and hazardous waste disposal facilities? What is the relationship between political and environmental patterns, and how does the disjunction between the two influence efforts to confront environmental degradation? Answering geographic questions of this sort requires careful analysis of the spatial character of pollution and the dynamic interactions between humans and their environment as a function of place.

As one example, during and shortly after the Manhattan Project,[1] workers at Los Alamos National Laboratory in northern New Mexico released some plutonium onto nearby canyon floors where it became attached to sedimentary particles. In subsequent decades, natural processes moved some of the sediments and their attached plutonium to the Rio Grande River, raising concerns about environmental and health hazards (Graf, 1994). Geographic analysis of the flows of plutonium through the general river system revealed that, on an average annual basis, 90 percent of the plutonium moving through the system was from sources other than the laboratory discharges—for example, fallout from atomic testing (see Figure 2.1). During some critical years, however, the contribution of plutonium from the laboratory amounted to as much as 86 percent of the total. No matter what the source, however, analysis of the plutonium budget and flows showed that only half of the plutonium entering the river was being transported through the system. The other half was being stored in the river system itself, and with a half-life of 24,000 years remains as a potential hazard, particularly if it is concentrated at some point up the food chain.

The location of these stored hazardous materials is, in fact, controlled by the spatial mechanics of the river system. In the northern Rio Grande River, plutonium absorbed into sediment is most often stored in floodplain deposits, channel fills, and reservoir sediments close to the point of its injection into the river system. Concentrations of plutonium in these sedimentary deposits are one to two orders of magnitude greater than concentrations in active channel sediments. In this way, understanding specific and localized geomorphic processes allowed this environmental hazard to be pinpointed, thereby improving risk assessment and mitigation measures.

ETHNIC CONFLICT

During the past two decades, ethnic conflict has undermined the existing social and political orders of many cities, countries, and world regions. Conflicts

---

[1] The Manhattan Project was a crash effort during the World War II to develop the first atomic weapons.

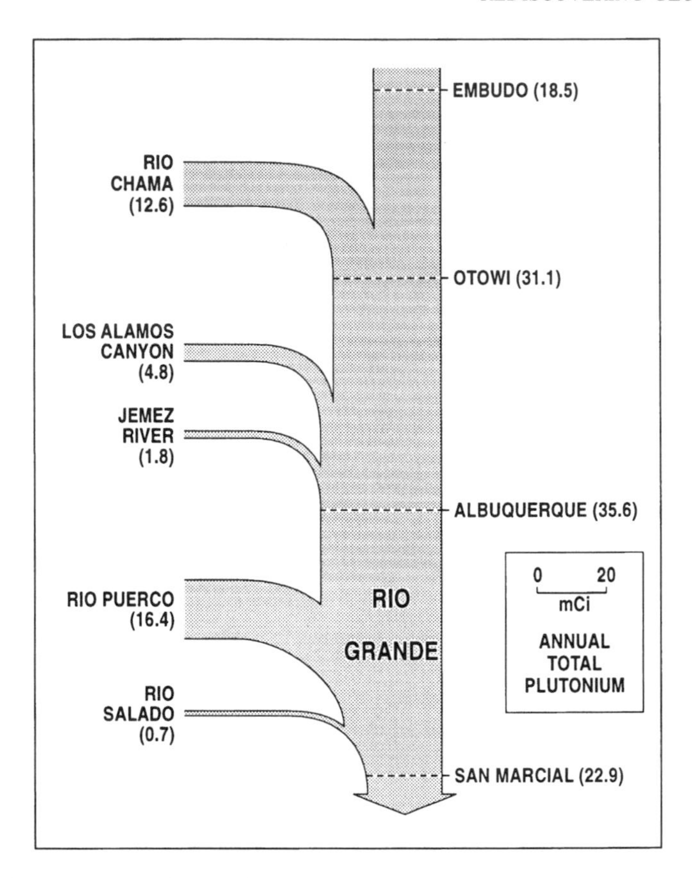

FIGURE 2.1 The flow of plutonium attached to sediment and dissolved in water in the northern Rio Grande River system in northern New Mexico. Annual contributions of plutonium to the river are mostly from fallout and industrial contributions from Los Alamos Canyon. The width of the arrow corresponds to the magnitude of plutonium transfer through the river system. The arrow representing flows through the main stem of the Rio Grande decreases in size in the downstream direction, indicating that plutonium is being stored in the river system. Source: Graf (1994).

between ethnic groups are manifest at a variety of scales, and in some cases they are precipitating major humanitarian crises. Consequently, ethnic conflict has increasingly attracted the attention of the scientific and policy-making communities. Efforts are being made to understand the causes and consequences of ethnic conflict, and policy makers are grappling with ways of mitigating intergroup hostilities.

Serious research on ethnic conflict has been hindered by the tendency on the part of many academics and policy makers to focus largely on individual nations and states. This very tendency shows why geography is so critical to the study of ethnic conflict. People look at and approach the world based on particular—often unacknowledged and untested—understandings of how it should be organized and territorially delimited. In the absence of any systematic analysis of those geographic understandings, the geography of ethnic conflict can easily be reduced simply to an exercise in naming the regions in which groups are located.

A serious geographic analysis of ethnic conflict can shed light on the spatial, territorial, and environmental dimensions of ethnic group interaction. It raises questions about the nature and significance of particular political-territorial structures, the role of boundaries, the character of flows between places of influence and control, and the role of the physical environment in shaping conflict and cooperation. Geographic work along these lines has clear implications for developing policy responses to ethnic conflicts. More broadly, it focuses attention on issues that are fundamental to an understanding of the dynamics of ethnic conflict, including the degree of legitimacy accorded particular territorial arrangements by different populations, the ways in which economic and social arrangements are at odds with dominant territorial structures, the implications of territorial arrangements for intergroup relations and understandings, and the effects of regional inequalities on political and social stability.

The insights to be gained from a geographic perspective on ethnic conflict can be illustrated by the geographic analysis of the Vance-Owen partition plan for Bosnia after the disintegration of Yugoslavia in the early 1990s (Jordan, 1993). The Vance-Owen plan came out of an attempt to divide the country on the basis of highly generalized ethnolinguistic maps. Analysis of daily commuting patterns (see Figure 2.2) showed, however, that the territorial units on which the Vance-Owen plan was based bore no relationship to the social and economic organization of Bosnia prior to the outbreak of conflict, which helped to explain why the Vance-Owen plan was so strongly opposed by those living there. In addition, analyses by geographers in the U.S. Department of State pointed out that by defining a large number of ethnic enclaves the Vance-Owen plan would result in an enormous amount of boundary length between opposing groups. When adversaries are not committed to peace, increasing boundaries between them may not be a promising avenue for conflict resolution. Taking into account geographic considerations of this sort is critical if policy analysts are to contribute to the resolution of complex disputes such as the one in Bosnia.

HEALTH CARE

How can society provide for the health needs of an aging population in the face of escalating costs, increasing dependence on publicly provided services, and tightening public-sector fiscal constraints? What responses are needed to help curtail the spread of AIDS (acquired immune deficiency syndrome)? How can society meet the needs of those who cannot afford adequate health care? Questions of this sort have attracted considerable attention of late in both academic

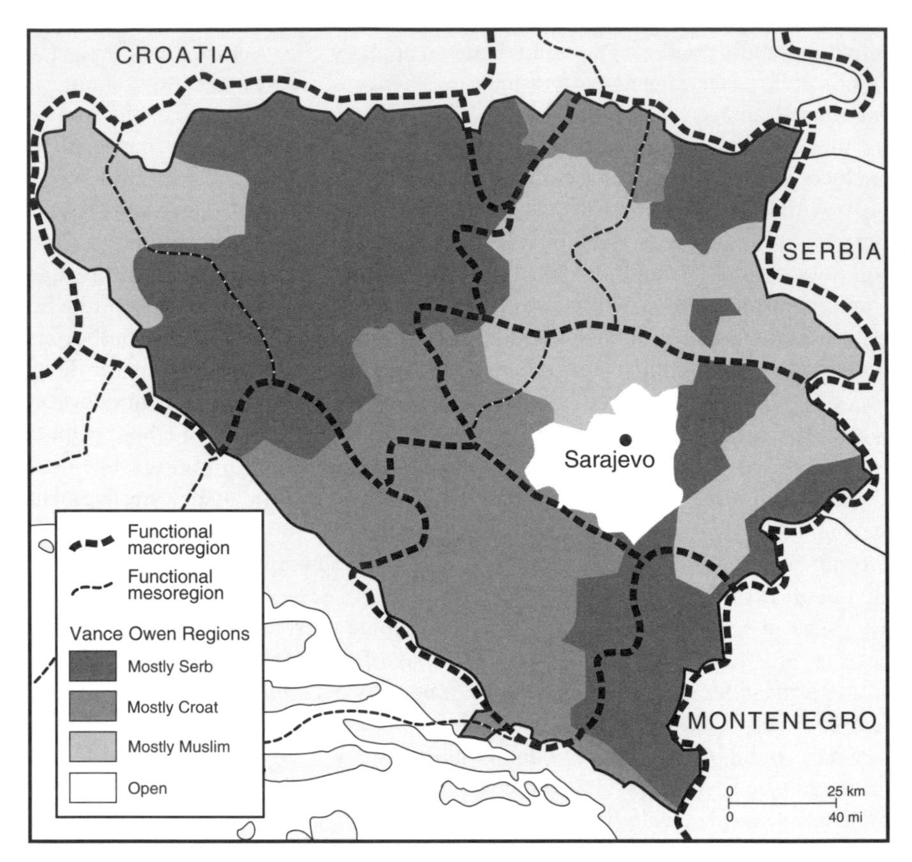

FIGURE 2.2 Relationship of preconflict functional regions (indicated by broken lines), determined by analysis of characteristics such as commuting patterns, to the Vance-Owen plan for partitioning Bosnia and Herzegovina (shaded areas), which was based on highly generalized ethnolinguistic maps. Source: After Jordan (1993).

and policy-making arenas. Indeed, such questions have assumed a sense of urgency as concerns have grown about the cost of, and equitable access to, health care services.

Geography has an important role to play in addressing such questions. Health care services are provided in particular places; effective decisions about where a particular service should be located must take into consideration the spatial organization of people, health problems, and related services. By focusing attention on locational efficiencies, a geographic analysis can point to specific ways of providing needed health care services cost effectively and, in many instances, can point to better ways of providing critical health services.

An example of the application of geographic perspectives to health care concerns low-birth-weight babies. Low-birth-weight children often have health problems that reduce their quality of life and that are costly to treat. Preventive health measures to reduce the incidence of low-birth-weight infants are therefore socially desirable and helpful in reducing future health care costs. One analysis in Iowa (Armstrong et al., 1991) showed that mothers who lived far from the hospital where their child was delivered were more likely to have a low-birth-weight infant. Why would we find this particular relationship? A map of Iowa (see Figure 2.3) shows the locations of hospitals that have more than 75 births per year. Large areas of the state are more than 20 miles from such a hospital. At the time of this study in 1990, it was the practice of the state to provide financial assistance for maternal and child health services only to hospitals with more than 75 births per year. This geographic analysis led to a review of that practice to consider supporting some smaller, rural hospitals that were strategically located to serve women in more remote areas. After 1990, the Iowa State Department of Public Health made state-supported nutrition, nursing, and maternal education services more accessible to pregnant women in the state by expanding Medicaid requirements so that a larger proportion of women became eligible to receive such services.

Beyond the question of infrastructure provision, geographic analysis has much to contribute to an understanding of the spread of disease. Ever since the source of a cholera epidemic was identified by mapping the distribution of cholera cases in nineteenth-century London, geographic analysis has been an important

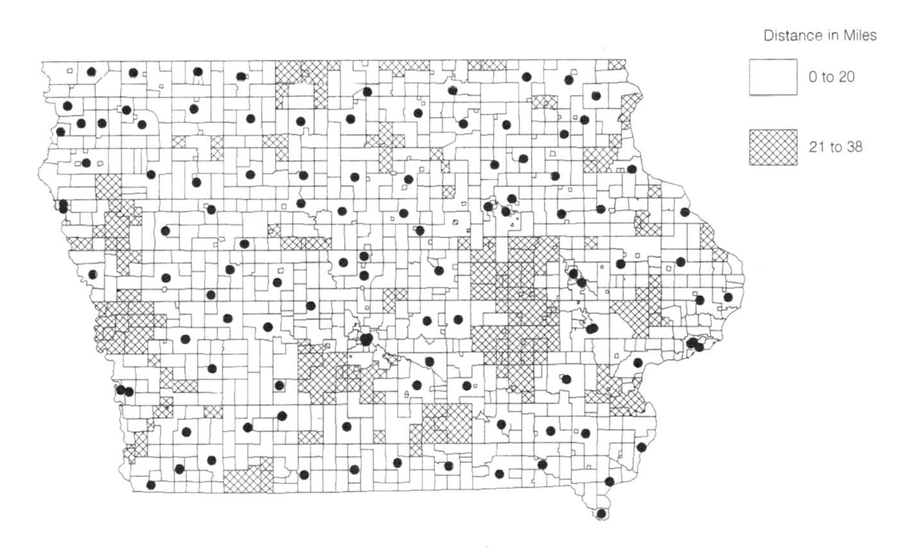

FIGURE 2.3 Map of Iowa showing distances to hospitals having more than 75 births per year. Lines represent the boundaries of the five-digit ZIP codes. Locations of hospitals are denoted by closed circles. Distances in miles are indicated by cross-hachuring. Source: Armstrong et al. (1991).

component of epidemiology. With the rise of new virulent viruses, the importance of a geographic perspective on infectious diseases is more critical than ever. Studies of the diffusion of AIDS (e.g., Gould, 1993; see Sidebar 5.9) offer the promise of enhancing our understanding of not only the behavior of the human immunodeficiency virus (HIV) but also the social and political conditions that have been most conducive to the spread of the virus.

GLOBAL CLIMATE CHANGE

During the summer of 1993, record rainfall brought devastating floods to the American Midwest. Plagued by drought only a few years earlier, California experienced damaging floods during the winter of 1994/1995. The summer of 1995 produced record heat waves throughout the United States and an unusually large number of tropical hurricanes. Are these weather events harbingers of long-term climate change that many experts predict based on their assessments of changes in the concentration of ''greenhouse gases'' in the Earth's atmosphere? Do they portend more frequent climate-related disasters than in the past?

Addressing these questions requires an understanding of the nature and dynamics of climate change. Climate change involves enormously complex interactions among the atmosphere, hydrosphere, and biosphere (see Figure 2.4). These interactions vary significantly across spatial scales. Thus, geographic perspectives that consider place and scale are essential for understanding potential climate changes. For example, geographers have been leaders in contributing to our understanding of large-scale climate patterns, especially those associated with the hydrologic cycle. As one instance, geographic research has shown that considerably more precipitation reaches the Earth's surface than most previous estimates suggest—and many climate models would indicate (Willmott and Legates, 1991).

One important facet of understanding global climate change is appreciating the nature of climatic variations since the last glacial maximum. By mapping past climate variations, identifying regional continuities, and focusing on the spatial relationships between climate and vegetation patterns, geographic analysis contributes to the larger interdisciplinary efforts to understand the operation of the climate system—past, present, and future. These contributions, in turn, are critical to the development of numerical models that are needed if scientists are to understand the extent to which humans may be modifying the climate system and the implications of those modifications.

COHMAP—the Cooperative Holocene Mapping Project—is an example of a recent interdisciplinary climate change research project with a strong geographic component (COHMAP, 1988; Wright et al., 1993). The simulations developed by COHMAP showed how variations in macroscale controls of climate—for example, the size of ice sheets, ocean temperatures, composition of the atmosphere, and the latitudinal and seasonal distributions of solar radiation—govern regional patterns of climate change (see Figure 2.5).

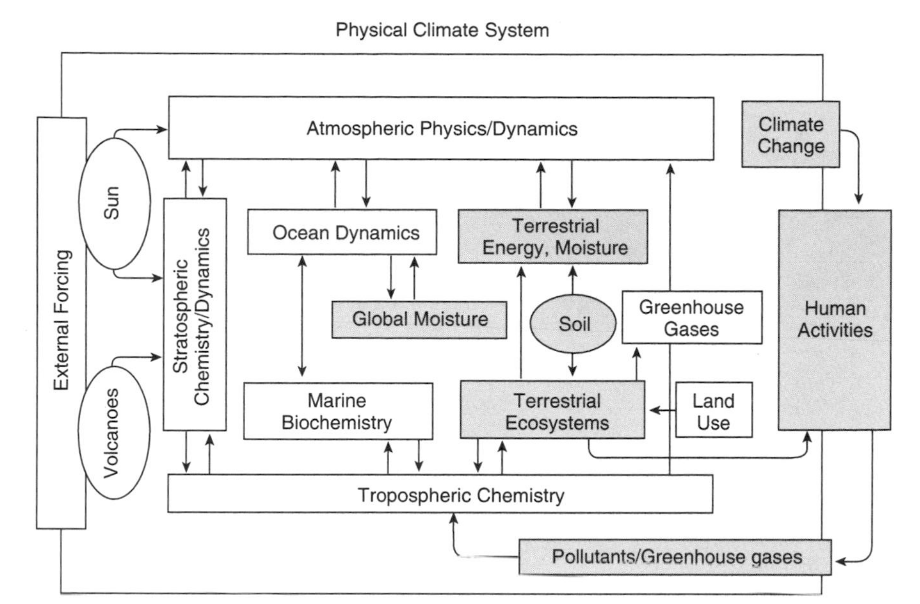

FIGURE 2.4 Revised version of the International Geosphere-Biosphere Programme model of the global climate system showing the forcing effects of human activities. Geographic research plays an important role in understanding the subsystems indicated by the shaded boxes. Source: After Williamson (1992).

COHMAP illustrates the kind of understanding that follows from an explicit geographic component within a larger interdisciplinary climate change study (Root and Schneider, 1995). Mapping and spatial analysis were essential in expressing and comparing the results of the COHMAP simulations and syntheses (Wright and Bartlein, 1993).

EDUCATION

One of the greatest challenges facing American society in the late twentieth century concerns education. The needs to improve the skills of the labor force and to meet the challenges of democratic citizenship in a fast-changing, increasingly complex world present enormous educational challenges (U.S. Department of Education, 1992; U.S. Department of Labor, 1991). What do tomorrow's citizens need to know to function effectively in a world characterized by both a globalized economy and changing local circumstances? What should schools be teaching students who may well hold several different kinds of jobs during the course of their lives? What educational experiences can promote personal enrichment in an age of television, telecommunications, computers, and hypermobility?

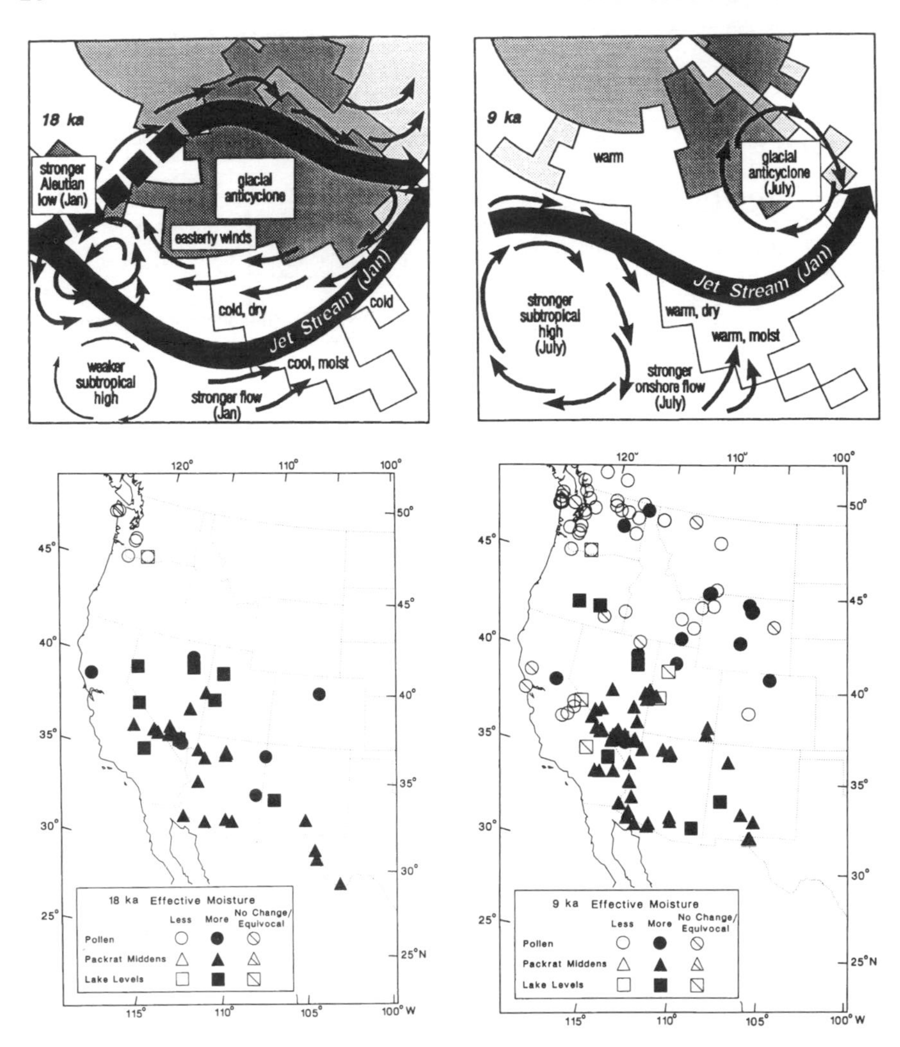

FIGURE 2.5 Climate-model simulations and paleoclimate observations for the western United States (after Thompson et al., 1993). The top panels summarize in a schematic fashion the results of the simulation of climate at 18 ka and 9 ka (ka = thousands of radiocarbon years ago) by using a general circulation model (the National Center for Atmospheric Research's Community Climate Model). The bottom panels show the pattern of effective moisture as inferred from fossil pollen evidence (circles), plant macrofossils from pack-rat middens (triangles), and geomorphic evidence of former lakes (squares).

It is clear that geography must be a part of any serious effort to meet the educational challenges implicit in these questions. Students need to be exposed to ideas and perspectives that cut across the physical-human divide, that consider how developments in one place influence those in other places, that focus attention on the ways in which local circumstances affect understandings and activities, and that foster an appreciation for the diversity of peoples and landscapes that comprise the Earth's surface. Recent outcries over the lack of geography in school curricula (see Chapter 1) reflect a growing recognition that an understanding of such matters is essential if the students of today are to function effectively in the world of tomorrow.

In response to demands for more and better instruction in geography, a set of voluntary national standards for geography education at the kindergarten through grade 12 (K–12) levels has been developed by a coalition of geographers and other educators (Geography Education Standards Project, 1994). In addition, geography alliances have been formed by the National Geographic Society in all 50 states to help school teachers become more effective geography instructors. The College Board is also adding a course and examination in geography to its Advanced Placement Program. These initiatives reflect an understanding that geography is not a luxury in a school curriculum. Instead, it is a necessary component of any reform initiative aimed at preparing students for the challenges of the twenty-first century.

In 1989 the Bush administration convened an education summit of the nation's governors at which they agreed that new goals needed to be established for American education. They determined that teaching and learning at the K–12 level should focus on a limited number of specific core subjects, including geography. Ultimately, national education goals were incorporated into legislation—the Educate America Act, which became public law in 1994. The act specifically included geography as a core subject, not only because geographic literacy was deemed to be important but also because geography instruction would be a vehicle for increasing classroom attention to contemporary issues and for integrating the content and skills associated with other core subjects.

CONCLUSION

Given the fundamental geographic underpinnings of so many critical issues facing society today, there is a clear need for an assessment of the role of geography in contemporary America. To provide such an assessment, this report turns first to a consideration of the discipline's perspectives and techniques (Chapters 3 and 4). This is followed by an examination of the relevance of the discipline in the scientific and policy-making arenas (Chapters 5 and 6). The report then concludes with a discussion of the challenges facing the discipline (Chapter 7) and the adjustments that are needed (Chapter 8) if geography is to respond to the demands being placed on it by scientists, policy makers, educators, and the private sector.

3

Geography's Perspectives

Geography's relevance to science and society arises from a distinctive and integrating set of perspectives through which geographers view the world around them. This chapter conveys a sense of what is meant by a geographic perspective, whether it be applied in research, teaching, or practice. Due to space limitations, it does not attempt to cite the many excellent examples of research illustrating geography's perspectives; the citations refer mainly to broad-ranging summaries of geographic research that are intended as resources for further reading.

Taking time to understand geography's perspectives is important because geography can be difficult to place within the family of academic disciplines. Just as all phenomena exist in time and thus have a history, they also exist in space and have a geography. Geography and history are therefore central to understanding our world and have been identified as core subjects in American education. Clearly, this kind of focus tends to cut across the boundaries of other natural and social science disciplines. Consequently, geography is sometimes viewed by those unfamiliar with the discipline as a collection of disparate specialties with no central core or coherence.

What holds most disciplines together, however, is a distinctive and coherent set of perspectives through which the world is analyzed. Like other academic disciplines, geography has a well-developed set of perspectives:

1. *geography's way of looking at the world* through the lenses of place, space, and scale;

2. *geography's domains of synthesis:*[1] environmental-societal dynamics relating human action to the physical environment, environmental dynamics linking physical systems, and human-societal dynamics linking economic, social, and political systems; and

3. *spatial representation* using visual, verbal, mathematical, digital, and cognitive approaches.

These three perspectives can be represented as dimensions of a matrix of geographic inquiry as shown in Figure 3.1.

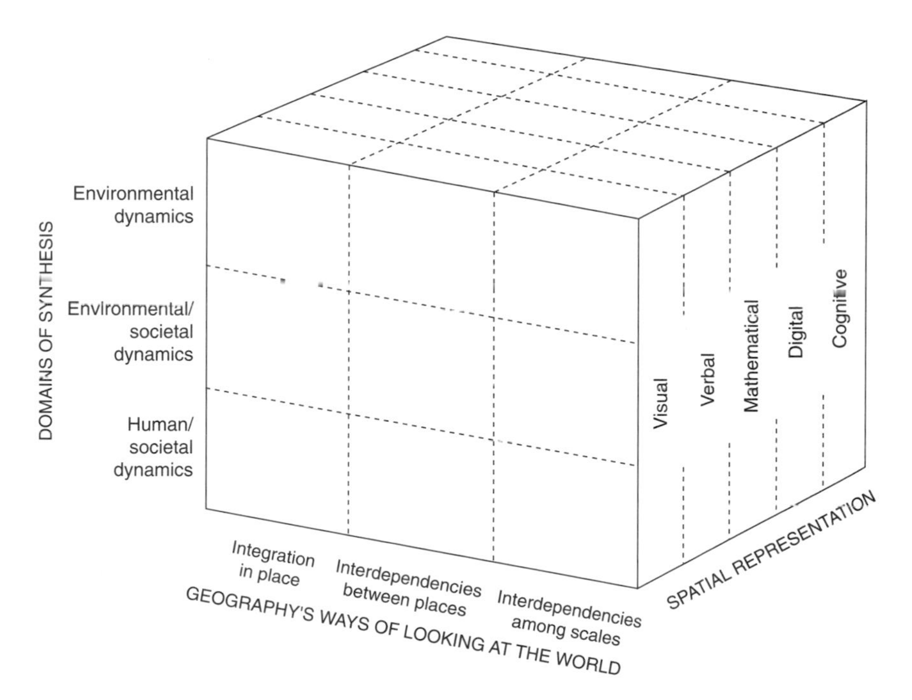

FIGURE 3.1 The matrix of geographic perspectives. Geography's ways of looking at the world—through its focus on place and scale (horizontal axis)—cuts across its three domains of synthesis: human-societal dynamics, environmental dynamics, and environmental-societal dynamics (vertical axis). Spatial representation, the third dimension of the matrix, underpins and sometimes drives research in other branches of geography.

[1] The term *synthesis,* as used in this report, refers to the way in which geographers often attempt to transcend the boundaries traditionally separating the various natural sciences, social sciences, and humanities disciplines in order to provide a broad-ranging analysis of selected phenomena. Such research benefits not only from bringing into one analysis ideas that are often treated separately in other disciplines but also from critically examining the disjunctures and contradictions among the ways in which different disciplines examine identical phenomena.

GEOGRAPHY'S WAYS OF LOOKING AT THE WORLD

A central tenet of geography is that "location matters" for understanding a wide variety of processes and phenomena. Indeed, geography's focus on location provides a cross-cutting way of looking at processes and phenomena that other disciplines tend to treat in isolation. Geographers focus on "real-world" relationships and dependencies among the phenomena and processes that give character to any location or *place*. Geographers also seek to understand relationships among places: for example, flows of peoples, goods, and ideas that reinforce differentiation or enhance similarities. Geographers study the "vertical" integration of characteristics that define place as well as the "horizontal" connections between places. Geographers also focus on the importance of scale (in both space and time) in these relationships. The study of these relationships has enabled geographers to pay attention to complexities of places and processes that are frequently treated in the abstract by other disciplines.

Integration in Place

Places are natural laboratories for the study of complex relationships among processes and phenomena. Geography has a long tradition of attempting to understand how different processes and phenomena interact in regions and localities, including an understanding of how these interactions give places their distinctive character.

The systematic analysis of social, economic, political, and environmental processes operating in a place provides an integrated understanding of its distinctiveness or character. The pioneering work of Hägerstrand (1970), for example, showed how the daily activity patterns of people can be understood as the outcome of a process in which individuals are constrained by the availability and geographic accessibility of locations with which they can interact. Research in this tradition since has shown that the temporal and spatial sequences of actions of individuals follow typical patterns in particular types of environments and that many of the distinctive characteristics of places result from an intersection of behavioral sequences constrained by spatial accessibility to the opportunities for interaction. Such systematic analysis is particularly central to regional and human geography, and it is a theme to which much geographic research continually returns. When such systematic analysis is applied to many different places, an understanding of *geographic variability* emerges. Of course, a full analysis of geographic variability must take account of processes that cross the boundaries of places, linking them to one another, and also of scale.

Interdependencies Between Places

Geographers recognize that a "place" is defined not only by its internal characteristics but also by the flows of people, materials (e.g., manufactured goods, pollutants), and ideas from other places. These flows introduce interdependencies between places that can either reinforce or reduce differences. For example, very different agricultural land-use practices have evolved under identical local environmental conditions as a result of the distance to market affecting the profitability of crops. At a macroscale, the widespread and global flow of Western cultural values and economic systems has served to reduce differences among many peoples of the world. An important focus of geography is on understanding these flows and how they affect place.

The challenge of analyzing the flows and their impacts on place is considerable. Such relationships have all the characteristics of complex nonlinear systems whose behavior is hard to represent or predict. These relationships are becoming increasingly important for science and decision making, as discussed in Chapters 5 and 6.

Interdependencies Among Scales

Geographers recognize that the scale of observation also matters for understanding geographic processes and phenomena at a place. Although geography is concerned with both spatial and temporal scales, the enduring dimension of the geographic perspective is the significance of *spatial* scales, from the global to the highly local.

Geographers have noted, for example, that changing the spatial scale of analysis can provide important insights into geographic processes and phenomena and into understanding how processes and phenomena at different scales are related. A long-standing concern of geographers has been the "regionalization problem," that is, the problem of demarcating contiguous regions with common geographic characteristics. Geographers recognize that the internal complexity and differentiation of geographic regions is scale dependent and, thus, that a particular set of regions is always an incomplete and possibly misleading representation of geographic variation.

Identifying the scales at which particular phenomena exhibit maximum variation provides important clues about the geographic, as well as the temporal, scope of the controlling mechanisms. For example, spectral analyses of temperature data, revealing the geographic scales at which there is maximum similarity in temperature, can provide important clues about the relative influence of microclimates, air masses, and global circulation on temperature patterns. A global rise in average temperature could have highly differentiated local impacts and may even produce cooling in certain localities because of the way in which global, regional, and local processes interact. By the same token, national and international economic and political developments can have highly differentiated impacts on the economic competitiveness of cities and states. The focus on scale enables geographers to analyze the impact of global changes on local events—and the impact of local events on global changes.

DOMAINS OF SYNTHESIS[2]

Geography's most radical departure from conventional disciplinary specializations can be seen in its fundamental concern for how humans use and modify the biological and physical environment (the *biophysical environment*) that sustains life, or *environmental-societal dynamics*. There are two other important domains of synthesis within geography as well: work examining interrelationships among different biophysical processes, or *environmental dynamics*, and work synthesizing economic, political, social, and cultural mechanisms, or *human-societal dynamics*. These domains cut across and draw from the concerns about place embedded in geography's way of looking at the world.

Environmental-Societal Dynamics

This branch of the discipline reflects, perhaps, geography's longest-standing concern and is thus heir to a rich intellectual tradition. The relationships that it studies—the dynamics relating society and its biophysical environment—today are not only a core element of geography but are also of increasingly urgent concern to other disciplines, decision makers, and the public. Although the work of geographers in this domain is too varied for easy classification, it includes three broad but overlapping fields of research: human use of and impacts on the environment, impacts on humankind of environmental change, and human perceptions of and responses to environmental change.

Human Use of and Impacts on the Environment

Human actions unavoidably modify or transform nature; in fact, they are often intended specifically to do so. These impacts of human action have been so extensive and profound that it is now difficult to speak of a "natural" environment. Geographers have contributed to at least three major global inventories of human impacts on the environment (Thomas, 1956; Turner et al., 1990; Mather and Sdasyuk, 1991) and have contributed to the literature of assessment, prescription, and argument regarding their significance. Studies at local and regional levels have clarified specific instances of human-induced landscape transformation: for example, environmental degradation in the Himalayas, patterns and processes of deforestation in the Philippines and the Amazon, desiccation of the Aral Sea, degradation of landscapes in China, and the magnitude and character of pre-Hispanic environmental change in the Americas.

Geographers study the ways in which society exploits and, in doing so,

---

[2]Citations in this section do not refer to major research contributions since these are the focus of Chapter 5. They refer the reader to books and articles that provide a more detailed discussion of the topic than can be provided here.

degrades, maintains, improves, or redefines its natural resource base. Geographers ask why individuals and groups manipulate the environment and natural resources in the ways they do (Grossman, 1984; Hecht and Cockburn, 1989). They have examined arguments about the roles of carrying capacity and population pressures in environmental degradation, and they have paid close attention to the ways in which different cultures perceive and use their environments (Butzer, 1992). They have devoted considerable attention to the role of political-economic institutions, structures, and inequities in environmental use and alteration, while taking care to resist portraying the environment as an empty stage on which social conflicts are acted out (Grossman, 1984; Zimmerer, 1991; Carney, 1993).

Environmental Impacts on Humankind

Consequences for humankind of change in the biophysical environment—whether endogenous or human induced—are also a traditional concern for geographers. For instance, geographers were instrumental in extending the approaches of environmental impact analysis to climate. They have produced important studies of the impact of natural climate variation and projected human-induced global warming on vulnerable regions, global food supply, and hunger. They have studied the impacts of a variety of other natural and environmental phenomena, from floods and droughts to disease and nuclear radiation releases (Watts, 1983; Kates et al., 1985; Parry et al., 1988; Mortimore, 1989; Cutter, 1993). These works have generally focused on the differing vulnerabilities of individuals, groups, and geographic areas, demonstrating that environmental change alone is insufficient to understand human impacts. Rather, these impacts are articulated through societal structures that give meaning and value to change and determine in large part the responses taken.

Human Perceptions of and Responses to Environmental Change

Geographers have long-recognized that human-environment relations are greatly influenced not just by particular activities or technologies but also by the very ideas and attitudes that different societies hold about the environment. Some of geography's most influential contributions have documented the roots and character of particular environmental views (Glacken, 1967; Tuan, 1974). Geographers have also recognized that the impacts of environmental change on human populations can be strongly mitigated or even prevented by human action. Accurate perception of change and its consequences is a key component in successful mitigation strategies. Geographers studying hazards have made important contributions to understanding how perceptions of risk vary from reality (Tuan, 1974) and how communication of risk can amplify or dampen risk signals (Palm, 1990; Kasperson and Stallen, 1991).

Accurate perceptions of available mitigation strategies is an important aspect of this domain, captured by Gilbert F. White's geographic concept of the ''range of choice,'' which has been applied to inform policy by illuminating the options available to different actors at different levels (Reuss, 1993). In the case of floodplain occupancy, for instance, such options include building flood control works, controlling development in flood-prone areas, and allowing affected individuals to absorb the costs of disaster. In the case of global climate change, options range from curtailing greenhouse gas (e.g., carbon dioxide) emissions to pursuing business as usual and adapting to change if and when it occurs. Geographers have assembled case studies of societal responses to a wide variety of environmental challenges as analogs for those posed by climate and other environmental change and have examined the ways in which various societies and communities interpret the environments in questions (Jackson, 1984; Demeritt, 1994; Earle, 1996).

Environmental Dynamics

Geographers often approach the study of environmental dynamics from the vantage point of natural science (Mather and Sdasyuk, 1991). Society and its roles in the environment remain a major theme, but human activity is analyzed as one of many interrelated mechanisms of environmental variability or change. Efforts to understand the feedbacks among environmental processes, including human activities, also are central to the geographic study of environmental dynamics (Terjung, 1982). As in the other natural sciences, advancing theory remains an overarching theme, and empirical verification continues to be a major criterion on which efficacy is judged.

Physical geography has evolved into a number of overlapping subfields, although the three major subdivisions are biogeography, climatology, and geomorphology (Gaile and Willmott, 1989). Those who identify more with one subfield than with the others, however, typically use the findings and perspectives from the others to inform their research and teaching. This can be attributed to physical geographers' integrative and cross-cutting traditions of investigation, as well as to their shared natural science perspective (Mather and Sdasyuk, 1991). Boundaries between the subfields, in turn, are somewhat blurred. Biogeographers, for example, often consider the spatial dynamics[3] of climate, soils, and topography when they investigate the changing distributions of plants and animals, whereas climatologists frequently take into account the influences that landscape heterogeneity and change exert on climate. Geomorphologists also account for climatic forcing and vegetation dynamics on erosional and depositional processes. The three major

[3]The term *spatial dynamics* refers to the movement, translocation of, or change in phenomena (both natural and human) over geographic space. The study of spatial dynamics focuses on the natural, social, economic, cultural, and historical factors that control or condition these movements and translocations.

subfields of physical geography, in other words, not only share a natural science perspective but differ simply with respect to emphasis. Each subfield, however, will be summarized separately here in deference to tradition.

Biogeography

Biogeography is the study of the distributions of organisms at various spatial and temporal scales, as well as the processes that produce these distribution patterns. Biogeography lies at the intersection of several different fields and is practiced by both geographers and biologists. In American and British geography departments, biogeography is closely allied with ecology.

Geographers specializing in biogeography investigate spatial patterns and dynamics of individual plant and animal taxa and the communities and ecosystems in which they occur, in relation to both natural and anthropogenic processes. This research is carried out at local to regional spatial scales. It focuses on the spatial characteristics of taxa or communities as revealed by fieldwork and/or the analysis of remotely sensed images. This research also focuses on historic changes in the spatial characteristics of taxa or communities as reconstructed, for example, from land survey records, photographs, age structures of populations, and other archival or field evidence. Biogeographers also reconstruct prehistoric and prehuman plant and animal communities using paleoecological techniques such as pollen analysis of lake sediments or faunal analysis of midden or cave deposits. This research has made important contributions to understanding the spatial and temporal dynamics of biotic communities as influenced by historic and prehistoric human activity as well as by natural variability and change.

Climatology

Geographic climatologists are interested primarily in describing and explaining the spatial and temporal variability of the heat and moisture states of the Earth's surface, especially its land surfaces. Their approaches are quite varied, including (1) numerical modeling of energy and mass fluxes from the land surface to the atmosphere; (2) in situ measurements of mass and energy fluxes, especially in human-modified environments; (3) description and evaluation of climatically relevant characteristics of the land surface, often through the use of satellite observations; and (4) the statistical decomposition and categorization of weather data. Geographic climatologists have made numerous contributions to our understanding of urban and regional climatic systems, and they are beginning to examine macroscale climatic change as well. They have also examined the statistical relationships among weather, climate, and sociological data. Such analyses have suggested some intriguing associations, for example, between urban growth and warming (Oke, 1979) and the seasonal heating cycle and crime frequency (Harries et al., 1984).

Geomorphology

Geomorphological research in geography emphasizes the analysis and prediction of Earth surface processes and forms. The Earth's surface is constantly being altered under the combined influences of human and natural factors. The work of moving ice, blowing wind, breaking waves, collapse and movement from the force of gravity, and especially flowing water sculptures a surface that is constantly being renewed through volcanic and tectonic activity.

Throughout most of the twentieth century, geomorphological research has focused on examining stability in the landscape and the equilibrium between the forces of erosion and construction. In the past two decades, however, emphasis has shifted toward efforts to characterize change and the dynamic behavior of surface systems. Whatever the emphasis, the method of analysis invariably involves the definition of flows of mass and energy through the surface system, and an evaluation or measurement of forces and resistance at work. This analysis is significant because if geomorphologists are to predict short-term, rapid changes (such as landslides, floods, or coastal erosion in storms) or long-term changes (such as erosion caused by land management or strip mining), the natural rates of change must first be understood.

Human-Societal Dynamics: From Location Theory to Social Theory

The third domain focuses on the geographic study of interrelated economic, social, political, and cultural processes. Geographers have sought a synthetic understanding of such processes through attention to two types of questions: (1) the ways in which those processes affect the evolution of particular places and (2) the influence of spatial arrangements and our understanding on those processes. Much of the early geographical work in this area emphasized locational decision making; spatial patterns and their evolution were explained largely in terms of the rational spatial choices of individual actors (e.g., Haggett et al., 1979; Berry and Parr, 1988).

Beginning with Harvey (1973), a new cohort of scholars began raising questions about the ways in which social structures condition individual behavior and, more recently, about the importance of political and cultural factors in social change (Jackson and Penrose, 1993). This has matured as an influential body of work founded in social theory, which has devoted considerable effort to understanding how space and place mediate the interrelations between individual actions and evolving economic, political, social, and cultural patterns and arrangements and how spatial configurations are themselves constructed through such processes (e.g., Gregory and Urry, 1985; Harvey, 1989; Soja, 1989; Wolch and Dear, 1989).

This research has gained wide recognition both inside and outside the discipline of geography; as a result, issues of space and place are now increasingly seen as central to social research. Indeed, one of the principal journals for interdisciplinary research in social theory, *Environment and Planning D: Society and Space*, was founded by geographers. The nature and impact of research that has sought to bridge the gap between social theory and conceptualizations of space and place are evident in recent studies of both the evolution of places and the interconnections among places.

Societal Synthesis in Place

Geographers who study societal processes in place have tended to focus on micro- or mesoscales. Research on cities has been a particularly influential area of research, showing how the internal spatial structure of urban areas depends on the operation of land markets, industrial and residential location decisions, population composition, forms of urban governance, cultural norms, and the various influences of social groups differentiated along lines of race, class, and gender. The impoverishment of central cities has been traced to economic, social, political, and cultural forces accelerating suburbanization and intraurban social polarization. Studies of urban and rural landscapes examine how the material environment reflects, and shapes, cultural and social developments, in work ranging from interpretations of the social meanings embedded in urban architecture to analyses of the impacts of highway systems on land uses and neighborhoods (Knox, 1994).

Researchers have also focused on the living conditions and economic prospects of different social and ethnic groups in particular cities, towns, and neighborhoods, with particular attention recently to how patterns of discrimination and employment access have influenced the activity patterns and residential choices of urban women (e.g., McDowell, 1993a, b). Researchers have also attempted to understand the economic, social, and political forces reinforcing the segregation of poor communities, as well as the persistence of segregation between certain racial and ethnic groups, irrespective of their socioeconomic status. A geographical perspective on such issues ensures that groups are not treated as undifferentiated wholes. By focusing attention on disadvantaged communities in inner cities, for example, geographers have offered significant evidence of what happens when jobs and wealthier members of a community leave to take advantage of better opportunities elsewhere (Urban Geography, 1991).

Geographical work on place is not limited to studies of contemporary phenomena. Geographers long have been concerned with the evolving character of places and regions, and geographers concerned with historical developments and processes have made important contributions to our understanding of places past and present. These contributions range from sweeping interpretations of the historical evolution of major regions (e.g., Meinig, 1986 et seq.) to analyses of the changing ethnic character of cities (Ward, 1971) to the role of capitalism in urban change (Harvey, 1985a, b). Studies along these lines go beyond traditional historical analysis to show how the geographical situation and character of places influence not only how those places develop but larger social and ideological formations as well.

Space, Scale, and Human-Societal Dynamics

Studies of the social consequences of linkages between places focus on a variety of scales. One body of research addresses spatial cognition and individual decision making and the impact of individual action on aggregate patterns. Geographers who study migration and residential choice behavior seek to account for the individual actions underlying the changing social structure of cities or shifting interurban populations. Research along these lines has provided a framework for modeling the geographical structure of interaction among places, resulting inter alia in the development of operational models of movement and settlement that are now widely used by urban and regional planners throughout Europe (Golledge and Timmermans, 1988).

Geographers also have contributed to the refinement of location theories that reflect actual private and public decision making. Initially, much of this research looked at locational issues at particular moments in time. Work by Morrill (1981) on political redistricting, for example, provided insights into the many ways in which administrative boundary drawing reflects and shapes political ideas and practices. More recent work has focused on the evolution of industrial complexes and settlement systems. This work has combined the insights of location theory with studies of individual and institutional behavior in space (Macmillan, 1989). At the interurban and regional scales, geographers have studied nationwide shifts in the location and agglomeration of industries and interurban migration patterns. These studies have revealed important factors shaping the growth prospects of cities and regions.

An interest in the relationship between individual behavior and broader-scale societal structures prompted geographers to consider how individual decisions are influenced by, and affect, societal structures and institutions (e.g., Peet and Thrift, 1989). Studies have tackled issues ranging from human reproduction and migration decisions to recreation and political protest. Researchers have shown how movement decisions depend on social and political barriers, the distribution of economic and political resources and broader-scale processes of societal restructuring. They have examined how the increased mobility of jobs and investment opportunities have affected local development strategies and the distribution of public resources between firms and households.

Indeed, there is new interest in theorizing the geographical scales at which different processes are constituted and the relationship between societal processes operating at different scales (Smith, 1992; Leitner and Delaney, 1996). Geographers recognize that social differences from place to place reflect not only differences in the characteristics of individual localities but also differences in how they are affected by societal processes operating at larger scales. Research has shown, for example, that the changing growth prospects of American cities and regions cannot adequately be understood without taking into account the changing position of the United States in the global system and the impact of this change on national political and economic trends (Peet, 1987; Smith and Feagin, 1987).

Geographic research also has focused explicitly on the spatial manifestations of institutional behavior, notably that of large multilocational firms; national, state, and local governments; and labor unions. Research on multilocational firms has examined their spatial organization, their use of geographical strategies of branch-plant location and marketing in order to expand into or maintain geographically defined markets, and the way their actions affect the development possibilities of different places (Scott, 1988b; Dicken, 1992). Research into state institutions has focused on such issues as territorial integration and fragmentation; evolving differences in the responsibilities and powers exercised by state institutions at different geographical scales; and political and economic rivalries between territories, including their impact on political boundaries and on geopolitical spheres of influence. Observed shifts in the location of political influence and responsibility away from traditional national territories to both local states and supranational institutions demonstrate the importance of studying political institutions across a range of geographical scales (Taylor, 1993).

SPATIAL REPRESENTATION

The importance of spatial representation as a third dimension of geography's perspectives (see Figure 3.1) is perhaps best exemplified by the long and close association of cartography with geography (see Chapter 4). Research emphasizing spatial representation complements, underpins, and sometimes drives research in other branches of geography and follows directly from the thesis that location matters. Geographers involved in spatial representation research use concepts and methods from many other disciplines and interact with colleagues in those fields, including computer science, statistics, mathematics, geodesy, civil engineering, cognitive science, formal logic, cognitive psychology, semiotics, and linguistics. The goals of this research are to produce a unified approach to spatial representation and to devise practical tools for representing the complexities of the world and for facilitating the synthesis of diverse kinds of information and diverse perspectives.

How geographers represent geographic space, what spatial information is represented, and what space means in an age of advanced computer and telecommunications technology are critical to geography and to society. Research linking cartographic theory with philosophies of science and social theory has demonstrated that the way problems are framed, and the tools that are used to structure and manipulate data, can facilitate investigation of particular categories of problems and, at the same time, prevent other categories of problems from even being recognized as such. By dictating what matters, representations help shape what scientists think and how they interpret their data (Sack, 1986; Harley, 1988; Wood, 1992).

Geographic approaches to spatial representation are closely linked to a set of core spatial concepts (including location, region, distribution, spatial interaction, scale, and change) that implicitly constrain and shape how geographers represent what they observe. In effect, these concepts become a priori assumptions underlying geographic perspectives and shaping decisions by geographers about how to represent their data and what they choose to represent.

Geographers approach spatial representation in a number of ways to study space and place at a variety of scales. Tangible representations of geographic space may be visual, verbal, mathematical, digital, cognitive, or some combination of these. Reliance on representation is of particular importance when geographic research addresses intangible phenomena (e.g., atmospheric temperature or average income) at scales beyond the experiential (national to global) and for times in the past or future. Tangible representations (and links among them) also provide a framework within which synthesis can take place. Geographers also study cognitive spatial representations—for example, mental models of geographic environments—in an effort to understand how knowledge of the environment influences peoples' behavior in that environment and make use of this knowledge of cognitive representation in developing approaches to other forms of representation.

Visual representation of geographic space through maps was a cornerstone of geographic inquiry long before its formal recognition as an academic area of research, yet conventional maps are not the only visual form used in geographic research. Figure 3.2 shows that conventional maps occupy a midpoint along a continuum of visual representation forms. This continuum can be defined by a dimension scale, which ranges from atomic to cosmological, and abstractness level, which ranges from images to line drawings.

Due to the centrality of geographic maps as a means for spatial representation, however, concepts developed for mapping have had an impact on all forms of spatial representation. This role as a model and catalyst for visual representation throughout the sciences is clear in Hall's (1992) recent popular account of mapping as a research tool used throughout science, as well as the recognition by computer scientists that maps are a fundamental source of many concepts used in scientific visualization (Collins, 1993).

An active field of geographic research on spatial representation involves formalizing the "language" for visual geographic representation. Another important field of research involves improved depiction of the Earth's surface. A notable example is the recent advance in matching computational techniques for terrain shading with digital elevation databases covering the conterminous United States (see Sidebar 3.1).

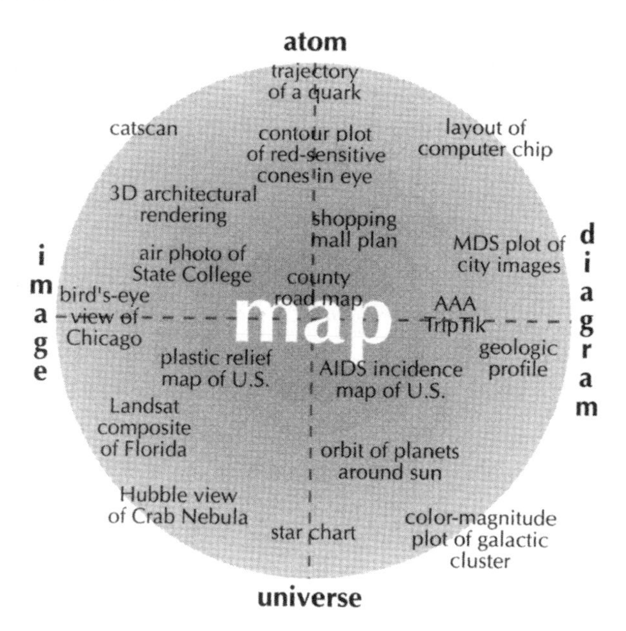

FIGURE 3.2 The conventional map is one of many visual representations of space used by geographers and other scientists. As one of a continuum of spatial representations, maps occupy a "fuzzy" category defined by an "abstractness level" (horizontal axis) and a "scale dimension" (vertical axis). Source: After MacEachren (1995, Figure 4.3).

Verbal representation refers to attempts to evoke landscapes through a carefully constructed description in words. Some of the geographers who have become best known outside the discipline rely almost exclusively on this form of representation. Geographers have drawn new attention to the power of both verbal and visual representations, exploring the premise that every representation has multiple, potentially hidden, and perhaps duplicitous, meanings (Gregory, 1994).

A current field of research linking verbal and visual forms of spatial representation concerns hypermedia documents designed for both research and instructional applications. The concept of a geographic script (analogous to a movie script) has been proposed as a strategy for leading people through a complex web of maps, graphics, pictures, and descriptions developed to provide information about a particular issue (Monmonier, 1992).

Mathematical representations include models of space, which emphasize location, regions, and distributions; models of functional association; and models of process, which emphasize spatial interaction and change in place. Visual maps, of course, are grounded in mathematical models of space, and it can be demonstrated that all map depictions of geographic position are, in essence, mathematical transformations from the Earth to the plane surface of the page or

SIDEBAR 3.1 Shaded Relief Map of the United States

U.S. Geological Survey researchers have produced what has been described as the most dramatic terrain depiction since Raisz's physiographic diagrams were published in 1940 (Lewis, 1992; see Figure 3.3). In creating this terrain map, Pike and Thelin (1989) developed procedures to integrate digital elevation models (DEMs) derived by digitizing 1:250,000 scale Defense Mapping Agency topographic maps. The digitized data were transformed to the Albers Equal Area Conic projection, which is appropriate for mapping the entire United States, and resampled to a resolution suitable for terrain shading at a scale of 1:3,500,000. Analytical hill-shading techniques were then applied, yielding three separate layers of information that were merged to produce the final map. The national digital terrain database behind the Pike and Thelin map has served as a catalyst for a number of other representational advances. Among them are systems for generating a sensation of depth in two-dimensional views (Moellering, 1989; Eyton, 1990) and for depicting the aspect of slope (Moellering and Kimerling, 1990; see Plate 1).

FIGURE 3.3. The Pike and Thelin shaded relief map of the conterminous United States. Source: Pike and Thelin (1989).

computer display screen. The combination of visual and mathematical representation draws on advantages inherent in each (see Plate 2).

A good example of the link between mathematical and visual representation is provided by the Global Demography Project (Tobler et al., 1995). In this project more than 19,000 digitized administrative polygons and associated population counts covering the entire world were extrapolated to 1994 and then converted to spherical cells. The data are available as a raster map, accessible on the World Wide Web from the National Aeronautics and Space Administration's Consortium for the International Earth Science Information Network, Socioeconomic and Economic Data Center, which supported the project.

Cognitive representation is the way individuals mentally represent information about their environment. Human cognitive representations of space have been studied in geography for more than 25 years. They range from attempts to derive "mental maps" of residential desirability to assessing ways in which knowledge of spatial position is mentally organized, the mechanisms through which this knowledge expands with behavior in environments, and the ways in which environmental knowledge can be used to support behavior in space. The resulting wealth of knowledge about spatial cognition is now being linked with visual and digital forms of spatial representation. This link is critical in such research fields as designing interfaces for geographic information systems (GISs) and developing structures for digital geographic databases. Recent efforts to apply the approaches of cognitive science to modeling human spatial decision making have opened promising research avenues related to way finding, spatial choice, and the development of GIS-based spatial decision support systems. In addition, research about how children at various stages of cognitive development cope with maps and other forms of spatial representation is a key component in efforts to improve geography education.

Digital representation is perhaps the most active and influential focus of representational research because of the widespread use of GISs and computer mapping. Geographers have played a central role in the development of the representational schemes underpinning GISs and computer mapping systems. Geographers working with mathematicians at the U.S. Census Bureau in the 1960s were among the first to recognize the benefits of topological structures for vector-based digital representations of spatial data. This vector-based approach (the Dual Independent Map Encoding system, more recently replaced by the Topologically Integrated Geographical Encoding and Referencing system, or TIGER) has become the linchpin of the Census Bureau's address-matching system. It has been adapted to computer mapping through an innovative system for linking topological and metrical geographic representations. Related work by geographers and other scientists at the U.S. Geological Survey's (USGS) National Mapping Division led to the development of a digital mapping system (the Digital Line Graph format) and has allowed the USGS to become a major provider of digital spatial data.

Geographers working in GIS research have investigated new approaches to raster (grid-based) data structures. Raster structures are compatible with the structure of data in remote sensing images, which continue to be a significant source of input data for GIS and other geographic applications. Raster structures are also useful for overlying spatial data. Developments in vector and raster data structures have been linked through an integrated conceptual model that, in effect, is eliminating the raster-vector dichotomy (Peuquet, 1988).

U.S. geographers have also played a leading role in international collaboration directed at the generalization of digital representations (Buttenfield and McMaster, 1991). This research is particularly important because solutions to key generalization problems are required before the rapidly increasing array of digital georeferenced data can be integrated (through GISs) to support multiscale geographic analysis. Generalization in the digital realm has proved to be a difficult problem because different scales of analysis demand not only more or less detailed information but also different kinds of information represented in fundamentally different ways.

Increasingly, the aspects of spatial representation discussed above are being linked through digital representations. Transformations from one representation to another (e.g., from mathematical to visual) are now routinely done using a digital representation as the intermediate step. This reliance on digital representation as a framework for other forms of representation brings with it new questions concerning the impact of digital representation on the construction of geographic knowledge.

One recent outgrowth of the spatial representation traditions of geography is a multidisciplinary effort in geographic information science. This field emphasizes coordination and collaboration among the many disciplines for which geographic information and the rapidly emerging technologies associated with it are of central importance. The University Consortium for Geographic Information Science (UCGIS), a nonprofit organization of universities and other research institutions, was formed to facilitate this interdisciplinary effort. UCGIS is dedicated to advancing the understanding of geographic processes and spatial relationships through improved theory, methods, technology, and data.

GEOGRAPHIC EPISTEMOLOGIES[4]

This survey of geography's perspectives illustrates the variety of topics pursued by geography as a scientific discipline, broadly construed. The methods and approaches that geographers have used to generate knowledge and understanding of the world about them—that is, its epistemologies—are similarly broad. The post-World War II surge in theoretical and conceptual geography, work

[4]The term *epistemology* refers to the methods of knowledge acquisition.

that helped the discipline take its place alongside other social, environmental, and natural sciences at that time, was triggered by adoption of what has been termed a "positivist" epistemology during the quantitative revolution of the 1960s (Harvey, 1969). Extensive use is still made of this approach, especially in studying environmental dynamics but also in spatial analysis and representation. It is now recognized, however, that the practice of such research frequently diverges from the ideals of positivism. Many of these ideals—particularly those of value neutrality and of the objectivity of validating theories by hypothesis testing—are in fact unattainable (Cloke et al., 1991; Taaffe, 1993).

Recognition of such limitations has opened up an intense debate among geographers about the relative merits of a range of epistemologies that continue to enliven the field (Gregory, 1994). Of particular interest, at various points in this debate, have been the following:

1. Approaches stressing the role of political and economic structures in constraining the actions of human agents, drawing on structural, Marxist, and structurationist traditions of thought that emphasize the influence of frequently unobservable structures and mechanisms on individual actions and thereby on societal and human-environmental dynamics—carrying the implication that empirical tests cannot determine the validity of a theory (Harvey, 1982).
2. Realist approaches, which recognize the importance of higher-level conceptual structures but insist that theories be able to account for the very different observed outcomes that a process may engender in different places (Sayer, 1993).
3. Interpretive approaches, a traditional concern of cultural geography, which recognize that similar events can be given very different but equally valid interpretations, that these differences stem from the varying societal and geographical experiences and perspectives of analysts, and that it is necessary to take account of the values of the investigator rather than attempting to establish his or her objectivity (Buttimer, 1974; Tuan, 1976; Jackson, 1989).
4. Feminist approaches, which argue that much mainstream geography fails to acknowledge both a white masculine bias to its questions and perspectives and also a marginalization of womens' lives in its analysis (McDowell, 1993b; Rose, 1993).
5. Postmodernist or "countermodernist" approaches, which argue that all geographic phenomena are social constructions, that understandings of these are a consequence of societal values and norms and the particular experiences of individual investigators, and that any grand theory is suspect because it fails to recognize the contingent nature of all interpretation. It is argued that this has resulted in a "crisis of representation," that is, a situation in which the relative "accuracy" of any representation of the world becomes difficult to adjudicate (Keith and Pile, 1993). Feminist and postmodern scholars argue that it is necessary to incorporate a diverse group of subjects, researchers, and ways of knowing if the subject matter of geography is to embrace humankind.

Geographers debate the philosophical foundations of their research in ways similar to debates among other natural scientists, social scientists, and humanists, although with a particular emphasis on geographical views of the world and on representation. These debates have not been restricted to the philosophical realm but have had very practical consequences for substantive research, often resulting in contrasting theoretical interpretations of the same phenomenon. For example, neopositivist and structural accounts of the development of settlement systems have evolved through active engagement with one another, and debates about how to assess the environmental consequences of human action have ranged from quantitative cost-benefit calculations to attempts to compare and contrast instrumental with local and indigenous interpretations of the meaning and significance of nature. In subsequent chapters we have not attempted to mark these different perspectives, choosing instead to stress the phenomena studied rather than the approaches taken. We attempt selectively to include leading researchers from different perspectives working on a particular topic, to the extent that their work can be constituted as scientific in the broad sense that we use that term (see Sidebar 1.1).

While we recognize that different perspectives frequently lead to intense debates engaging very different views of the same phenomenon, there is no space in this report to detail these debates. Such often vigorous interchanges and differences strengthen geography as both a subject and a discipline, however, reminding researchers that different approaches may be relevant for different kinds of questions and that the selection of any approach shapes both the kind of research questions asked and the form the answers take, as well as the answers themselves.

4

Geography's Techniques

This chapter provides a brief discussion of contributions made by geographers to the development of techniques for observation, display, and analysis of geographic data. With respect to observation, the chapter addresses two extremes on the geographic scales of observation: local fieldwork and remote sensing. With respect to the display and analysis of data, the chapter examines cartography, visualization, geographic information systems (GISs), and spatial statistics.

The techniques that geographers use in their work are not developed in a vacuum. They are developed to address specific problems and, thus, reflect the focus of the discipline at particular times. These techniques reflect the conscious decisions of geographers about the kinds of information that are important to collect; the spatial scales at which information should be collected, compiled, analyzed, and displayed; data sampling strategies and experimental designs; data representation; and methods for data analysis. As theoretical paradigms change, so do the techniques for empirical research. Thus, advancement of the discipline goes hand in hand with the development of new and improved techniques for collecting, analyzing, and interpreting information. Sidebars 4.1 and 4.2 illustrate the close relationship between advancement of the discipline and technique development.

The variety of perspectives in geography and a recognition of how different world views and experiences influence theoretical work (see Chapter 3) help geographers remain conscious of the influence of theory on technique development (and the reverse). The current popularity of GISs, for example, both reflects and reinforces the influence of spatial analytic theories in the discipline. There

SIDEBAR 4.1 Potential Evapotranspiration

Until the 1940s, practical algorithms were unavailable for reliably evaluating the relative wetness or dryness of climates, relationships between precipitation and stream runoff, the amount of irrigation necessary to maximize crop yield, and a number of other hydroclimatic problems. What was missing was an easy-to-use, reliable, and yet physically realistic way of estimating the time-integrated atmospheric demand for land surface moisture. Surrogates for atmospheric moisture demand, such as pan evaporation and air temperature, were used out of necessity, but they were conceptually flawed and often produced highly biased estimates.

Working with colleagues at the Laboratory of Climatology in New Jersey during the 1930s and 1940s, C.W. Thornthwaite devised a relatively straightforward characterization of atmospheric moisture demand (termed "potential evapotranspiration," or E_0) and a practical means of estimating it (Wilm et al., 1944). Thornthwaite's contribution to climatological understanding has endured not only because his E_0 concept is physically grounded, but because relatively reliable estimates of E_0 can easily be made from measurements (or estimates) of monthly air temperature (T) and day length (h).

Thornthwaite's E_0 as well as his climatic water-budget algorithms (Thornthwaite and Mather, 1955) have found many varied uses, ranging from evaluating local hydroclimatic problems to assessing the geographic variability of evapotranspiration on regional, continental, and even global scales. His separation of E_0, E, and β, and representation of E as $E = E_0 \beta$—where E is the actual evapotranspiration and β is a dimensionless measure of land surface moisture conductance—in particular, suggests which environmental characteristics should be observed and estimated. Thornthwaite's conceptualization, in other words, has significantly advanced our understanding of how to sample the environment to assess the hydroclimate of a place or region.

is a lively debate about whether the popularity of GISs is hindering the development of other theoretical approaches, such as social theory, that require different techniques for empirical analysis.

This chapter illustrates some of the ways in which geographers have made substantial contributions to empirical scientific techniques through their methodological research. Some of these techniques were developed by other disciplines for other purposes and were adapted by geographers to meet the special challenges posed by the study of spatial and temporal aspects of phenomena, processes, and events. Some of the techniques developed by geographers have found widespread use in other disciplines and in the public and private sectors generally. Perhaps the best current examples are GISs that store, manipulate, and display geographically referenced information. The potential of GISs to handle large quantities of spatially related information fills an important need in research, education, and applied work in the public and private sectors. The geographic information system, in addition to being a stimulus for theoretical research in spatial representation (see

Chapter 3), is a major topic of techniques-related research in geography and it is one of the principal topics of specialization within the field, currently demanded by students and employers alike.

The focus of this chapter on techniques for empirical analysis should not be taken to mean that methodological contributions in geography have been restricted to observation and hypothesis testing. For the past 20 years the discipline has been a fertile field of theoretical research, particularly in conceptualizing and modeling geographic processes. In many ways these theoretical developments have given the discipline its secure intellectual foundation.

OBSERVATION

Observation of phenomena and events is central to geography's concern for accurately representing the complexity of the real world. The traditional and still widely practiced method of observation is through direct "on-the-ground" contact between geographer and subject through field observation and exploration. Fieldwork is particularly effective for making observations at micro- to meso-scales, as typified, for example, by the study of single watersheds or cities.

SIDEBAR 4.2 Spatial Diffusion and Epidemics

Theoretical work on the spatial diffusion of innovations began with the novel conceptual and methodological contributions of Torsten Hägerstrand (1953; see also Hägerstrand, 1967). Recognizing that the formal concepts of spatial structure, pattern, and random choice were all at work, he adopted a Monte Carlo approach (for the first time in any of the human sciences) based on simulation methods used by John Von Neumann and Stanislav Ulman to solve intractable mathematical problems encountered on the Manhattan Project during World War II. The diffusion of innovations through a human population was assumed to occur through information passing by interpersonal contact.

His theory formalized ideas that were developing at the time in cultural geography, and it has since been applied to a wide range of phenomena. For instance, research on the spread of contagious diseases from measles to AIDS has built on Hägerstrand's formalisms and has profoundly improved our understanding of epidemics. This research has shown that the waves of many epidemics that spread through populations depend to a considerable degree on the geographic distribution of interpersonal contact patterns (Cliff et al., 1986; see Figure 4.1).

More recently, it has been argued that spatial diffusion is influenced as much by the societal context within which diffusion occurs, as by the actions of individuals (Brown, 1981; Blaut, 1987). A conceptual shift to a "supply-side" perspective has changed the questions asked about spatial diffusion from a focus on susceptibility of individual adopters to a focus on the role of societal forces in the diffusion process, including the way in which such forces structure the space through which a pathogen might move. This has necessitated different kinds of information, and different empirical research strategies, from those characterizing earlier work.

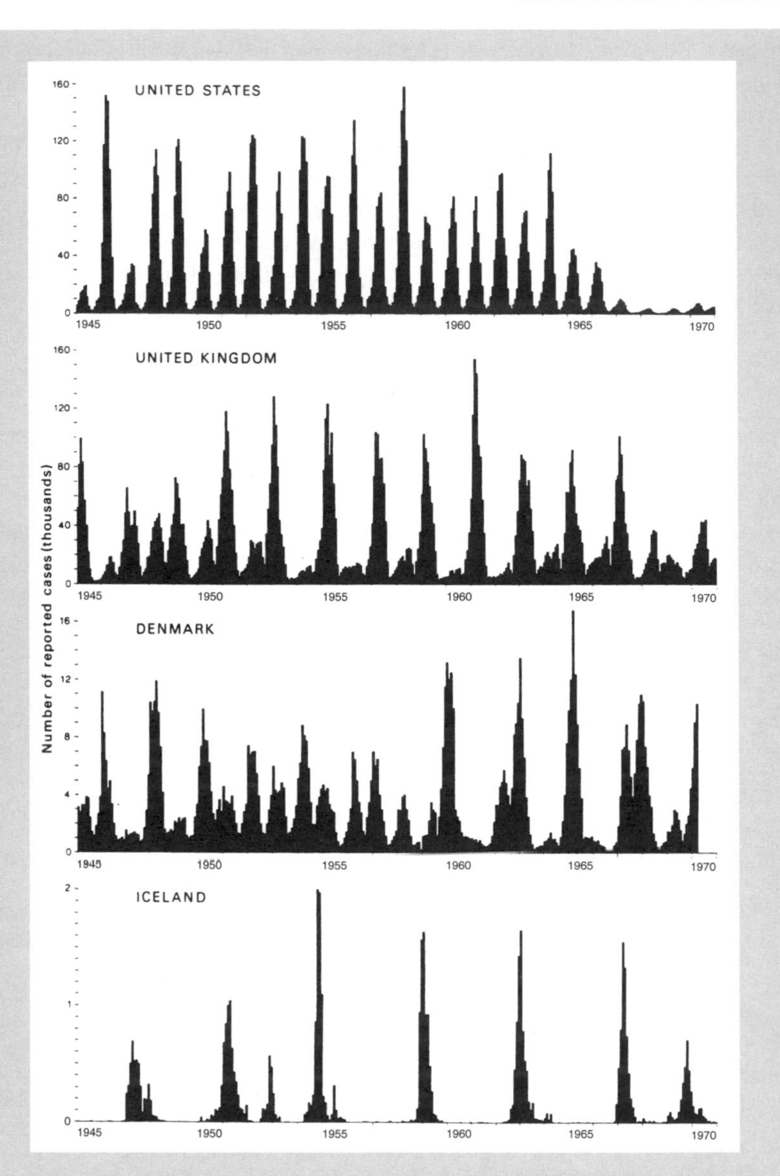

FIGURE 4.1 Measles epidemics occur in cycles, as illustrated here in plots showing the number of reported cases of measles per month in the United States, the United Kingdom, Denmark, and Iceland. The frequency of measles epidemics is related to the size and isolation of the affected populations. Epidemics tend to occur more often in large, well-connected, mobile populations (e.g., the United States) and less often in smaller, more isolated populations (e.g., Iceland). Source: Cliff et al. (1981).

Fieldwork is an intensive endeavor. It can require substantial investments of human and financial resources, particularly if carried out over extended periods of time.

The intensive nature of fieldwork makes it impractical for macroscale observations of the Earth's surface. Such observations are best made by using remote sensing techniques that utilize air- or spaceborne platforms and sensors. The development of these techniques, and especially the collection of remote sensing data, have often been led by geographers.

Field Observation/Exploration

The principal laboratory for geographic investigation is the field. Indeed, Sayer (1993) has argued that the intensive, comparative case-study research enabled by field observation is central to understanding the variations between places that geographers specialize in studying. Many of geography's most compelling questions center around changes in the physical and built landscape. Addressing those questions usually requires field observation and spatial sampling. Geographers interested in social patterns and processes also use archival research, interviewing and surveying techniques, and participant observation methods that are associated with the social sciences more generally.

Fieldwork allows geographers to make direct observations in places where local data are missing or unreliable and to check the validity of existing secondary sources such as census statistics. While the increased availability of remote sensing imagery would seem to reduce the need for fieldwork in some research, in truth it makes such work even more important because accurate interpretation of imagery depends on detailed knowledge of the actual patterns on the ground. To decide, for example, which areas on a forested scene represent healthy old-growth forest as opposed to disease-infested old-growth or young secondary forest, researchers need field data on the composition and distribution of different forest stands so that they can assess how such stands are "seen" by remote sensors.

Fieldwork may also be needed to test the validity of interpretations (e.g., a vegetation map) that others have made based on remote sensing imagery—just as it is necessary to test the validity of other secondary data sources, digital or otherwise. The wide availability of digitized secondary datasets (not just interpreted imagery but also census data and other information) makes it easy for students and researchers to download information and perform their own analyses. Unfortunately, such datasets frequently do not include detailed descriptions of the origins and reliability of the information. In cases where these "meta-data" are missing, the digital datasets may only be useful if the researcher is able to assess their reliability through fieldwork. In a digital age, when it seems so easy to collect observations "automatically" from space or secondhand from online datasets, fieldwork becomes more, rather than less, important to good scholarship.

The contributions of geography to the practice of field research derive from the discipline's emphasis on location and synthesis. As noted in Chapter 3, geographers are concerned with distributions and spatial patterns of phenomena. In these connections they have contributed to our understanding of distributions and patterns through the development of innovative field mapping techniques. To geographers, field maps are more than orientation aids. They are a tool to record and to uncover relationships among observations. Geographers have developed field mapping techniques to shed light on everything from spatial cognition to the origin and diffusion of cultural traits. With the explosion of GISs, global positioning systems (GPSs), and related technologies, geographers are at the forefront in automating the compilation, manipulation, and analysis of field observations (see Figure 4.2).

The field geographer's interest in distributions and spatial patterns is part of a larger concern with synthesis: how and why particular phenomena come together in specific places to create distinctive environments. This concern leads geographers in the field to observe and study a wide range of physical and social phenomena. Research on land reform, for example, might involve soil sampling as well as interviews with affected individuals.

The enduring importance of fieldwork in geography extends beyond research to pedagogy. At a time when new ways are being sought to promote environmental and cultural awareness through education, geographic fieldwork offers a unique and valuable perspective. Field excursions are incorporated routinely into many geography courses. They are designed to teach students about the environment in which they live and to encourage them to be inquisitive about the processes that shape landscapes and cultures. Fieldwork thus provides both a tool for the acquisition of knowledge and a means of promoting awareness and appreciation of culture and the environment.

Remote Sensing

Remote sensing is defined here as the detection and recording of electromagnetic radiation signals from the Earth's surface and atmosphere using sensors placed aboard aircraft and satellites. These signals are usually recorded in digital form, where each "digit" denotes one piece of information about an average property of a small area of the Earth.

Geographers have been using remote sensing data since they first became available about 30 years ago. Geographers who study the Earth's climate, for example, use satellites to collect data on atmospheric conditions for monitoring and predicting change. Remotely sensed data also are very useful in creating and updating maps of physical, biological, and cultural features at the Earth's surface. The ability of certain sensor systems to "see" through cloud cover, and their unrestricted access to all portions of the Earth, provide information that may not be available from other sources.

FIGURE 4.2 This photograph shows a GPS (mounted on the rider's back) being used to record positional data while traversing remote highlands of the Dominican Republic for a research project on Quaternary paleoclimatology and biogeography. Data were subsequently entered into a GIS to map trail routes and geomorphic features under study.

Geographers have played important roles in efforts to use satellite data for taking inventory of and monitoring land cover, both regionally and globally. For instance, a joint U.S. Geological Survey (EROS Data Center)/Center for Advanced Land Management Information Technologies project recently demonstrated a method using multisource data that successfully characterized 159 land cover classes for the United States (see Plate 11). The acquisition of global land cover data has been recognized as a top priority by the National Research Council and the International Geosphere-Biosphere Programme (e.g., Townshend, 1992; NRC, 1994).

Geographers have also played key roles in the collection and processing of remotely sensed data for the National Oceanic and Atmospheric Administration's Coastwatch Change Analysis Project, a joint state and federal program designed to monitor environmental changes in coastal wetlands, uplands, and submerged habitats. Klemas et al. (1993), for example, developed a coastal land cover classification system for use with satellite imagery. The system is compatible with existing coastal mapping programs and databases and with GIS. Dobson and co-workers (Dobson and Bright, 1991; Dobson et al., 1993) have conducted prototype studies in the Chesapeake Bay watershed and have helped develop a protocol for mapping locations, characteristics, and changes in coastal zone habitats. Jensen and co-workers (Jensen et al., 1993a, b) have used remote sensing data to predict the effect of sea-level changes on coastal wetlands (see Plate 3).

Scale is a fundamental issue for collection of data by remote methods. In this context, scale can refer to the spatial, spectral, radiometric, or temporal characteristics of sensor systems, all of which—singly and in combination—affect the quality and usefulness of the collected data. Geographers have made valuable contributions in understanding the spatial- and temporal-scale dependence of geographic data and in determining optimal measurement scales for remotely gathered data.

Geographers are also involved in developing new technologies for integrating, analyzing, and visualizing multiresolution satellite data in integrated geographic information systems. The integration and visualization of satellite imagery with other types of data hold enormous potential for researchers, resource managers, and decision makers as they strive to inventory, understand, and manage the Earth's human and natural environment.

Sampling and Choice of Observations

An important component of much geographic research is estimation of the values of variables through sampling. Evaluating the efficacy of sampling designs is an important topic of research in geography and an important aspect of applying geography's techniques.

Traditionally, sample collection in geography utilized sampling designs borrowed from classical statistics, but for many geographic data, classical sampling designs may not produce representative samples. Random samples are not necessarily representative samples because of the way processes can vary in time and space. For spatially referenced data there is no consistent relationship between the number of observations and their representativeness. Quite commonly, other information is available that permits the differential weighting of sample observations. Such Bayesian weighting schemes, for example, are becoming especially important in interpreting the geographic patterns of disease distributions. The "samples" of observed diseases are hypothesized to be drawn from known processes, and the interest is in observing differences between observed and expected patterns of disease given the geographic distribution of conditions known to affect the likelihood of the disease being present in a population (Langford, 1994). Generating samples from known processes, computing the reference distribution and finding the relationship of an observed sample estimate to it, is often accomplished using Monte Carlo simulation methods (Openshaw et al., 1987, 1988).

Many samples of geographic data are taken from datasets designed for other purposes (e.g., from the census). This resampling of samples confounds classical statistical probability assessments and hypothesis testing. A number of geographers have begun to move away from classical statistical approaches toward more flexible approaches that incorporate geographic understanding. The move from "direct" to "indirect" estimation techniques that rely on knowledge from related observations to estimate the conditions of small areas illustrates this change. In qualitative approaches in human geography, formal sampling designs are often questioned, since space is not conceived of as an empty space whose content is to be captured through systematic samples—but, rather, as a differentiated space with meanings attached to areas that change across space in noncontinuous ways. In qualitative analyses a contrast is made between time, which is seen as one-dimensional and unidirectional, and space, which is seen as multidimensional and ordered in many different ways. For the qualitative geographer, who is often cultivating the middle ground between the universality of science and the particularity of history, interpreting the meaning of change in space becomes the goal and purposive sampling the tool for this end.

An example of a new approach to sampling design and evaluation is the ongoing effort to assess global climate and climatic change from weather station measurements. Rain gauge networks provide spatial samples of the continually varying global precipitation field. Their spatial distributions are neither random by design nor demonstrably random in effect (see Figure 4.3); nevertheless, the nodes of these sampling networks (the rain gauge locations) are arguably the best representations of historical precipitation variability. Standard statistical approaches are inadequate for assessing precipitation variability from such samples for the reasons mentioned above. In their place, computer-intensive, nonparametric methods of evaluating the rain gauge networks and, in turn, the precipita-

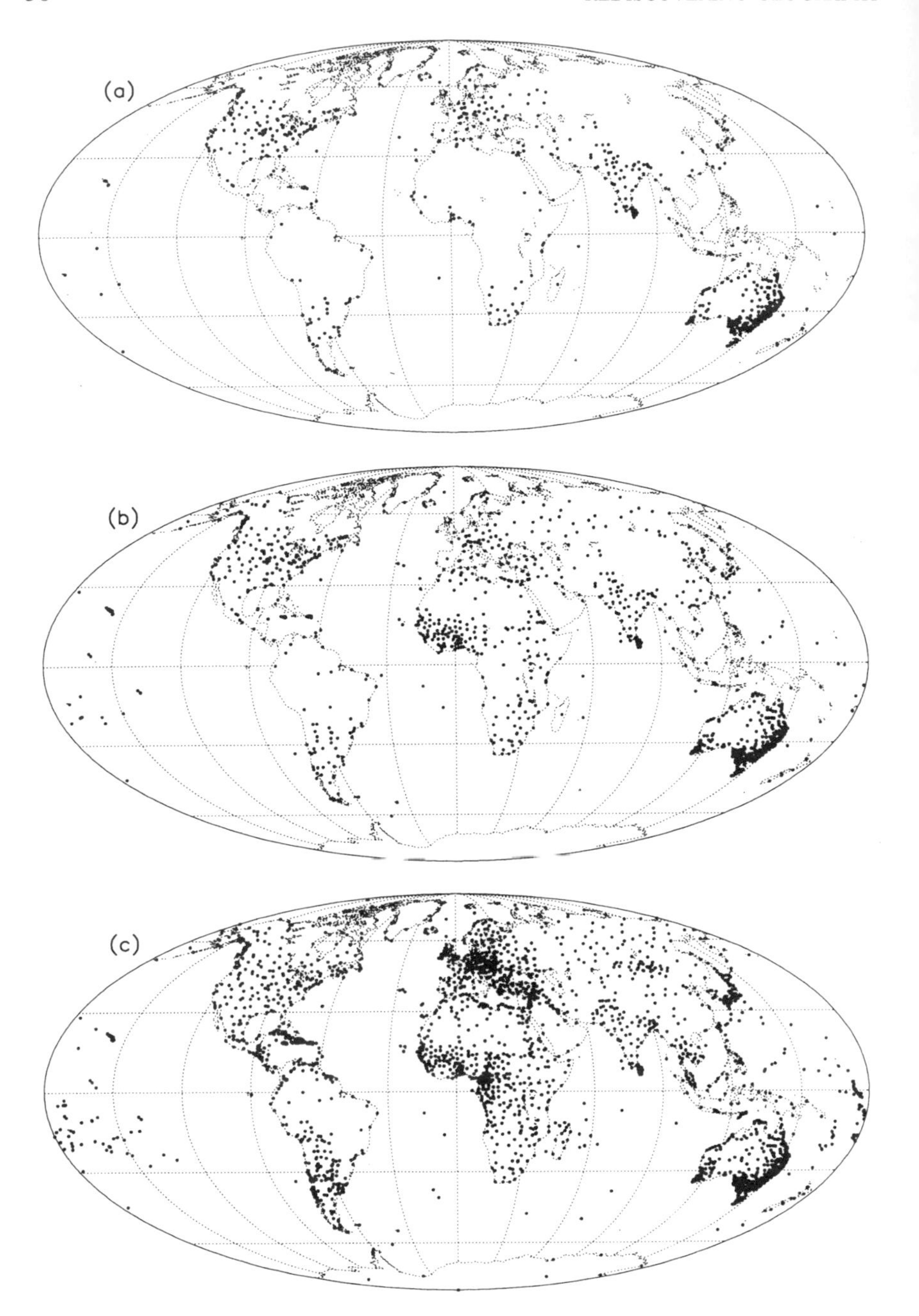

FIGURE 4.3 Spatial distribution of precipitation stations in the National Center for Atmospheric Research's World Surface Climatology for (a) 1900, (b) 1930, and (c) 1960. Station locations are denoted by closed circles. Source: Willmott et al. (1994).

tion fields that they represent have proven useful, when informed by climatological understanding.

DISPLAY AND ANALYSIS

The traditional tool in geography for the display of spatially referenced information is the map. *Cartography* is a subdiscipline traditionally concerned with formalized procedures for making maps. To many, the term *map* connotes a fixed, two-dimensional paper product containing point, line, and area data. During the past generation, however, advances in data collection, storage, analysis, and display have greatly expanded this traditional view. The "modern" map is a dynamic and multidimensional product that exists in digital form. The advent of such maps has opened up new fields of research and application for geographic investigation.

Any geographer educated 25 years ago who returned to the discipline today would be impressed by the methods geographers now use to record and process spatial information (Laurini and Thomas, 1992). The changes extend beyond the development of GISs to new techniques for geographic visualization and spatial statistical analysis, which provide for an increasingly complex and contextual understanding of the world. This same observer would also be impressed by the problems that remain to be solved. For example, a substantial methodology now exists for statistical analysis of spatial data, but it has not yet been integrated into GISs. Indeed, as a platform for the investigation of scientific questions, GISs are still in their infancy. Many geographers believe that a large dividend would come from integrating GISs as information science with visualization techniques and spatial analysis methods.

The following subsections provide a brief review of some of the substantive methodological contributions of the discipline to display and analysis techniques. These include cartography, GISs, geographic visualization, and spatial statistics.

Cartography

The traditional close association between geography and maps is appropriate given the discipline's concern with space and place. The symbiotic link between geographers and maps has ensured the persistence of cartography as a subdiscipline of geography within most academic settings.

The field of cartography has changed enormously during the past three decades, primarily because of the widespread availability of computers. Computers have made possible new forms of symbolization, such as dynamic (i.e., animated) maps, customized maps for individual users, and interactive maps. They have also made possible new methods for scientific visualization and spatial data analysis.

Geographic cartographers have made especially valuable contributions to the development of automated mapping systems. Their research on map reading processes, map production techniques, cartographic generalization, and cartographic design has facilitated the automation and formalization of what had been an intuitive manual procedure. With generalization, for example, a conceptual model has been devised that separates the subjective and holistic approaches of traditional cartography into discrete subcomponents that have been successfully incorporated into digital mapping software (McMaster and Shea, 1992). Cartographers have also worked to prevent the inadvertent misuse of computer mapping systems and maps by developing expert systems for map production.

Some of the most interesting and potentially useful research conducted by geographers today is in the realm of dynamic or animated cartography. Animation enables the visualization of changes in phenomena across space, through time, and in attributes of the phenomena themselves (see Sidebar 4.3). One of the earliest animated maps of the microcomputer era showed the spread of AIDS at the county level in Pennsylvania (Gould, 1989). This animation was used to highlight the initial concentrations and spatial diffusion of the disease more effectively than a sequence of static maps. The intention of this dramatic portrayal was to inform and educate health care researchers and the general population. The cartographic techniques developed in this research subsequently led to inclusion of a series of animated maps in one of the best-selling CD-ROM encyclopedias.

Geographers have led the way in research on another new cartographic format: electronic atlases and atlases on CD-ROMs. In a recent project undertaken jointly by Florida State University, the Florida Department of Education, and IBM, an atlas of Florida was published on a CD-ROM and distributed to all schools in the state (''Atlas of Florida,'' 1994; see Figure 4.5). This format permits the inclusion of multimedia material that could not be accommodated by traditional printed text: audio, video, animation, or a multitude of photographs and other graphics. Electronic atlases and related geographic programs are already proving to be effective in educational settings, especially at the kindergarten through grade 12 levels. The newly released ''ExplOregon: A Geographic Tour of Oregon'' (a 1995 multimedia CD-ROM developed by William Loy with Digital Chisel software by Pierian Spring Software) is changing the way that the geography of Oregon is taught and learned in schools throughout the state. As national standards for geography education are developed, educational aids such as electronic atlases will become indispensable.

Geographic Information Systems

Geographic information systems were defined in 1992 by the U.S. Geological Survey as ''computer system[s] capable of assembling, storing, manipulating, and displaying geographically referenced information'' (USGS, 1992). Such systems, in fact, have power, utility, and importance far beyond this definition, both within and beyond the field of geography. Their most valuable potential capability,

SIDEBAR 4.3 Climate and Vegetation Change

Map animation has been used to examine continental-scale changes in vegetation in response to climate change. Figure 4.4 shows a series of *isopoll* maps of the observed and simulated distribution of spruce pollen in eastern North America during the past 18,000 years. The observed pollen maps are based on data from 328 fossil pollen sites across the region. The simulated maps were generated by using general circulation models of past climates at 3,000-year intervals, together with response surfaces that describe the dependence of the modern distribution of pollen on climate (COHMAP, 1988). These individual "snapshot" maps were animated to explore the sometimes small differences between observed and simulated patterns of spruce and other pollen types and to study the changing patterns of all taxa at time scales shorter than those represented here. A combination of visual and statistical analyses of these and other data has been interpreted to support the *dynamic equilibrium hypothesis* that explains vegetation change as a response to continuous climatic forcing. Continental-scale vegetation patterns appear to reflect the changes of climate during the past 18,000 years with lags no greater than about 1,500 years (McDowell et al., 1991; Prentice et al., 1991; Webb and Bartlein, 1992).

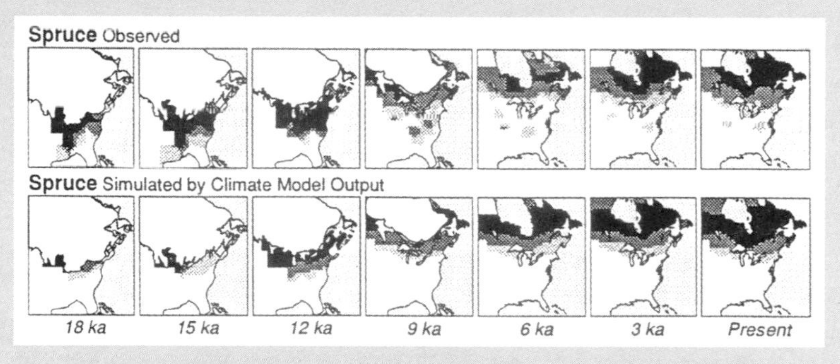

FIGURE 4.4 Isopoll maps showing the observed and simulated distribution of spruce (*Picea*) pollen in the eastern United States. The light, medium, and dark stippling represent pollen abundances greater than 1, 5, and 20 percent, respectively. The symbol "ka" denotes thousands of years before present. Source: Webb and Bartlein (1992, Figure 2).

which sets them apart from computer mapping systems, is the ability to perform spatial analyses to address research and application questions.

Fundamental to the successful propagation of GISs is the development of methods for representing and coding spatial data (see Sidebar 4.4). GISs can be used to perform an extensive variety of spatial operations and analyses on properly coded data. At the most elementary level are computations of distances, areas, centroids, gradients, and volumes. More complex operations that add spatial

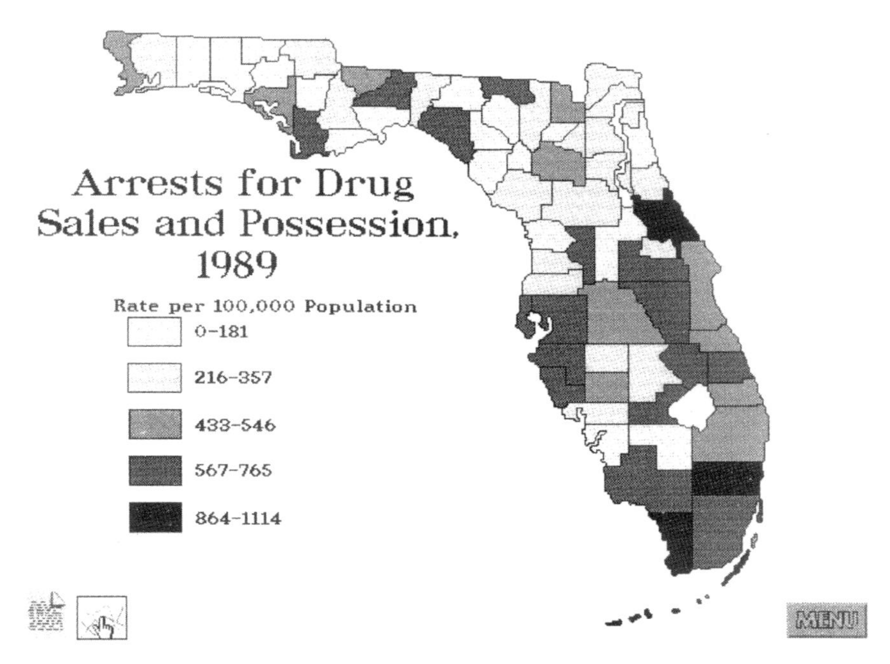

FIGURE 4.5 Choropleth map from the ''Atlas of Florida'' CD-ROM showing arrests for drug sales and possession by county for 1989. By clicking on the graph icon at the lower left of the screen, the user can display an animated bar graph that depicts changes over time in drug-related crimes and enforcement. Source: ''Atlas of Florida'' (1994).

referencing to a basic calculation are also possible—for example, going beyond questions about the total length of a city's sewer lines to questions about the total length of sewer lines in a given area in a particular city and what proportion of this length is more than 50 years old. GISs are also capable of more complicated operations such as (1) calculating new spatial datasets based on attributes of existing data—for example, calculating slopes from elevations; (2) comparing two or more spatial datasets based on user-specified criteria—for example, identifying toxic waste sites that are situated on permeable soil; (3) delimiting areas that possess certain characteristics defined by the user—for example, delimiting locations of commercially zoned land within 2 miles of an interstate highway; and (4) modeling the possible outcomes of alternative processes and policies—for example, determining the impact of flooding along the Mississippi River given the presence or absence of levees (see Plate 10).

GISs are being used to facilitate a variety of management and planning decisions in both the public and the private sectors. For instance, in Wake County, North Carolina, potential sites for schools, libraries, and other facilities have been determined by identifying adequately sized parcels of vacant land and providing information about utilities, topography, and demographic characteris-

SIDEBAR 4.4 Locational Coding

With the help of its Geography Division, the U.S. Bureau of the Census has developed and released into the public domain a digital data file describing roads, rivers, and many other features of the United States at a nominal scale of 1:100,000. Attribute information of these Topologically Integrated Geographical Encoding and Referencing (TIGER) system files includes the locations of street addresses on all defined road segments. Many software developers and vendors have written computer programs that allow users to read and display these files and to add other information containing street addresses. Street addresses have become the locational reference code from which near-absolute locations can be established with an accuracy of approximately 20 m. Through address matching, distances can be computed through the street network between, for example, patients and hospitals, the scene of crimes and police patrol routes, and the homes of children and schools.

New locational coding schemes now make it possible to customize information for specific applications. For example, businesses can use this information to better target their marketing efforts. Government can use this information to improve the delivery of social services or economic assistance.

Geographers who know how to access and manipulate TIGER data are in great demand by the many public and private organizations that have discovered the importance of thinking spatially in an increasingly geographically diverse world.

tics of the local population. By incorporating population data from the 1990 census into their GIS, Wake County planners can generate population projections, vacancy rate estimates, and growth rates in demands for services (Juhl, 1994).

Plate 4 provides a further illustration of the capabilities of GISs to integrate data from diverse sources to create products for planners and policy makers. The figure shows a map of a portion of Columbia, South Carolina, that was created by combining 1990 census block-level data on racial composition with the county assessor's parcel-level data on land use. Proportional pie graphs provide a visualization of the varying levels of racial integration within blocks. This type of display would be especially useful for political reapportionment.

Another challenge in which GISs have found use is in monitoring natural resources. For example, GISs are being used to assess the impact of water releases from the Glen Canyon Dam on water flow in the Colorado River through the Grand Canyon (Powers et al., 1994). Researchers constructed databases that contain spatial coordinates of the study area and "layers" of information about the canyon's vegetation, surficial geology, and hydrology. These three characteristics are being monitored and analyzed to determine trends through time in riparian vegetation growth, habitat, and "events" such as channel scouring and channel constriction.

In addition, GISs have a growing role in international policy and planning associated with human welfare. One particularly compelling example, cutting across the issues outlined in Chapter 2, involves the Bangladesh Flood Action Plan (FAP) of the International Council for Scientific Unions-International Geographical Union (ICSU-IGU). FAP is a major international research and policy development effort designed to provide flood warning, coordinate assistance efforts during flood events, and develop long-term flood mitigation plans. The project brings geographers in Bangladesh together with an international team of experts (coordinated by the IGU) to build a knowledge and technology base at the University of Dhaka. A particularly important component of the project involves rapid field mapping by local technicians using handheld GPS monitors integrated with pen computers. Data are gathered and fed directly into the GIS to allow for updated mapping and integration with remotely sensed images.

Although GISs are being utilized at ever-increasing rates, their full potential remains to be realized. Geographers have an intrinsic interest in GISs from at least three perspectives: (1) as users of GISs for research and applications; (2) as contributors to the development of GIS methods, theories, and applications; and (3) as educators. The interest in GISs as an education tool is becoming increasingly important because their applications are growing rapidly and their impacts promise to be powerful. Geographers will be responsible for preparing future generations of GIS users and must provide them with strong backgrounds in understanding geographic processes and patterns, spatial analysis, and spatial visualization techniques.

To the end user of a GIS, its operation can be a deceptively simple "black box" that generates answers to queries at the press of a few keys. There is an inherent danger in this apparent simplicity, however, in that users can easily and unknowingly misuse the power of GISs to produce irrelevant or erroneous solutions (Cartography and Geographic Information Systems, 1995). Users need considerable background knowledge of the subject matter to which GISs are being applied, as well as an understanding of the analytical operations available on the systems, in order to know what questions to ask, the relevant variables to invoke, and how to recognize nonsensical procedures and answers.

An outgrowth of GISs (linked to geographic visualization and spatial statistics—both discussed separately below) is the development of geographic information analysis (GIA) tools. Such tools typically use an existing GIS as a base (or rely on data structures originally designed to support GIS) to which numerical analysis and sophisticated visual display methods are linked. Among the more successful early GIA tools was the Geographical Analysis Machine (GAM), the goal of which was to generate an automated answer to the question of "where" to look for patterns by initially looking everywhere (Openshaw et al., 1987). GAM is part of a larger effort to develop a "computational human geography," an approach to human geography that builds on the massive databases of social and economic data being generated together with inductive approaches to arriving at useful generalizations. Application of the original GAM resulted in successful identification of clusters of leukemia cases. The method was, however, limited by its use of brute force (nonefficient) methods.

In a recent extension of GAM, Openshaw (1995) has implemented a space-time-attribute analysis machine (STAM) linked directly to GIS. The goal of STAM is to determine a set of GIS operators that, when applied to a database query, will identify cases that form a statistically unusual pattern. In its most recent implementation STAM incorporates the use of "genetic" algorithms (i.e., algorithms that adapt to their environment, for example, by defining the appropriate spatial scale and resolution for a particular problem context).

Another significant development in GIA is the incorporation of object-oriented programming (OOP) concepts to extract the data needed for an analysis from a GIS, solve spatial analytic problems, and subsequently link the solutions back to the GIS for display of results and further analysis. One example of this approach involves development of custom routing software for use by the U.S. Agency for International Development (USAID) in decision making related to food aid transport in southern Africa (Ralston, 1994). Southern Africa imports large quantities of food grains, and the state of the economy dictates that much of this is in the form of food aid. The software developed for USAID is based on an OOP approach linked to the commercial GIS, ArcInfo. The combination provides a flexible tool for prepositioning food in storage facilities, setting prices for acquisition of more carrying capacity, determining where to add more storage, making distribution decisions, choosing best modes and routes for transport, and determining the location and cost of bottlenecks. The OOP approach was found to be particularly useful in quickly adapting the decision model to deal with changes in obstacles to distribution.

Geographic Visualization

Geographic visualization (GVis) can be defined as "the use of concrete visual representations ... to make spatial contexts and problems visible, so as to engage the most powerful human information processing abilities, those associated with vision" (MacEachren et al., 1992, p. 101). The dramatic increase in volume of georeferenced data being collected and generated today is exceeding our capacity to analyze and digest it. Using the power of human vision to recognize patterns and synthesize spatial information increases the capacity of geographic researchers to cope with this data volume. For example, a simple 48 × 48 matrix of fiscal transfers for the United States generates 2,256 pieces of information for each time period considered. Such information can be concisely summarized in a simple yet effective visualization (e.g., see Figure 4.6).

GVis combines display with analysis capabilities to enable the search for patterns and relationships; the identification of anomalies; the analysis of directions and flows; the delineation of regions; and the integration of local, regional, and global information (see Figure 4.6). The development of flexible GVis tools

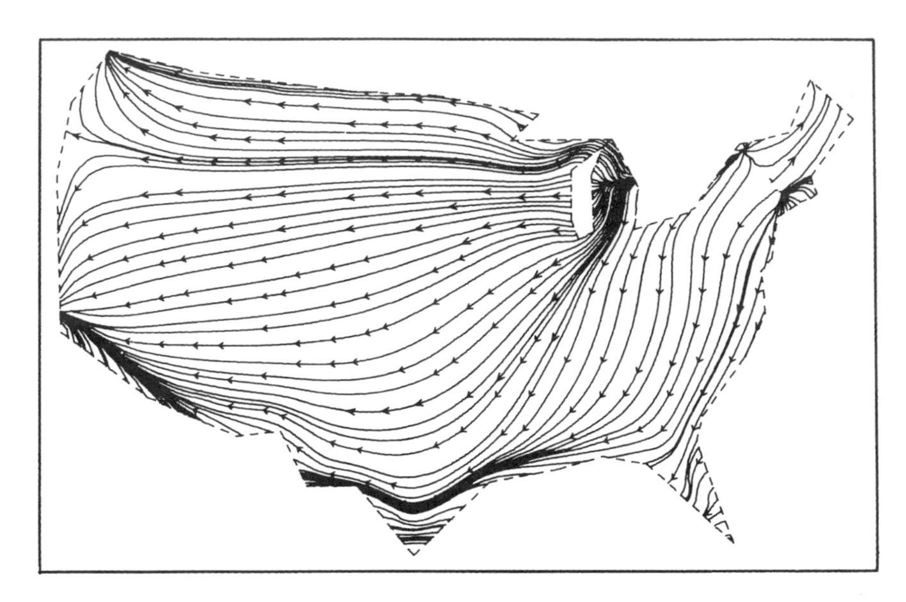

FIGURE 4.6 Estimated trajectories of fiscal transfers via federal accounts for 1975 (Tobler, 1981). The flow pattern has been estimated from a rather coarse 61 × 95 finite difference mesh, having an approximate resolution of 400 km. The flow lines depict the theoretical paths of fiscal transfers computed from the potential field obtained as the solution of Poisson's equation. This simple but effective visual depiction dramatically illustrates the general direction of flow and also depicts partitioning into flow regions quite clearly. The resolution upon which it was based, which was partly a function of computational power in 1981, does not support identification of particular origins and destinations.

is an important topic of geographic research because such tools are essential to fully exploit the information "content" of georeferenced data. Geography as a discipline is involved with GVis in three ways: development, application, and assessment of the implications of its use.

The most active topics of GVis research are exploratory spatial data analysis (ESDA) and the application of multimedia to spatial analysis, education, and policy decisions. Research on ESDA includes work on extending classical statistics to deal with spatial data, as described later in this chapter. Research in this field also includes the development of new data transformation and symbolization techniques and the development of computer interfaces to allow interactive analysis of spatial data (see Sidebar 4.5).

In the field of multimedia research, cartographers are developing techniques and computer interfaces that allow animation to be used as a tool for spatial pattern recognition. Multimedia tools are also being developed to link maps, graphics, text, and data to understand the complexities of geographic processes in problems involving human-environment interactions. GVis concepts also provide structure for the design of multimedia tools for interactive learning. Multimedia visualization technology is important in the context of digital geographic libraries. A particularly innovative development in this context is the first comprehensive collection of Native American maps (Andrews and Tilton, 1993). The library contains high-resolution images of original maps stored at a variety of resolutions and accessible through a comprehensive cross-referenced indexing tool that takes advantage of the nonlinear structure inherent in hypermedia applications.[1]

Geographers are exploiting the dynamic capabilities of GVis—its abilities to directly manipulate model parameters or map elements or its capacity for animation—to analyze complex spatial processes. As noted in Sidebar 4.3, for example, animation has been used to understand tree species migration across North America and to compare it with predictions from climate models.

Beyond its role in exploratory research, interactive GVis is rapidly evolving as a tool in policy formulation. A good example is Shiffer's (1993) multimedia environment, which was developed to allow participants at public meetings concerned with airport siting to hear the noise that would be generated at specified locations in relation to the proposed facility.

GVis and scientific visualization in general represent a shift away from strict quantitative analysis to increased reliance on qualitative sensory perception. Particularly when used with modeling and simulation, GVis has important ramifications for geography and the rest of science. These include fundamental issues of how research questions are framed and even what constitutes a problem worthy of investigation. As geographers move away from the use of fixed maps and avail themselves of the multiple perspectives permissible with GVis, they must come to grips with how the "truth" of the representations generated can be judged—and even what truth is. There is a growing need to address such issues as truth in representation, particularly in relation to the use of GVis and GISs in public policy applications.

Spatial Statistics

The analysis of geographically referenced information poses statistical challenges not faced in most other disciplines. First, observations are not always scalar numbers, such as points on a map. They may be multidimensional, consisting of lines, areas, and volumes. Second, the observations may be spatially or spatiotemporally covariant—that is, the values of observations made in one location may depend on the values of observations from other locations or from the same

[1]The term *hypermedia* is generally used to refer to data structures in which discrete packets of information (e.g., text, graphics, sound) are electronically cross-referenced and linked for easy and efficient access (e.g., Lindholm and Sarjakoski, 1994). A good example of a hypermedia application is the World Wide Web.

SIDEBAR 4.5 Interactive Analysis of Spatial Data

Among the more innovative techniques in exploratory spatial data analysis is *geographic brushing,* a computer procedure that dynamically links and displays scatterplots and maps of data (Monmonier, 1989; see Figure 4.7). The procedure allows analysts to highlight individual or groups of data points on a scatterplot displayed on a computer screen and to see where the points appear on other scatterplots or on the map. Alternatively, the analyst can highlight specific geographic regions on the map and see where the data from these regions fall on the scatterplots. A temporal "brush" allows the analyst to quickly step through a time series of maps and scatterplots to investigate changes in spatial relationships.

FIGURE 4.7 Sample screen from a computer display that illustrates the technique of *geographic brushing*. The scatterplots at the top of the screen show relationships among metropolitan population, per capita income, and percentage of homes with cable television for the United States. The bottom of the screen displays a slidebar (the *temporal brush*) that can be used to select the time period displayed on the scatterplots. In this case, the year selected is 1985. The analyst uses the *scatterplot brush* to select data on the scatterplot—in this case, the hachured region on the middle-left scatterplot is selected. States with data that plot in this region are highlighted (shaded) on the *geographic brush* in the middle of the screen. Source: Monmonier (1989).

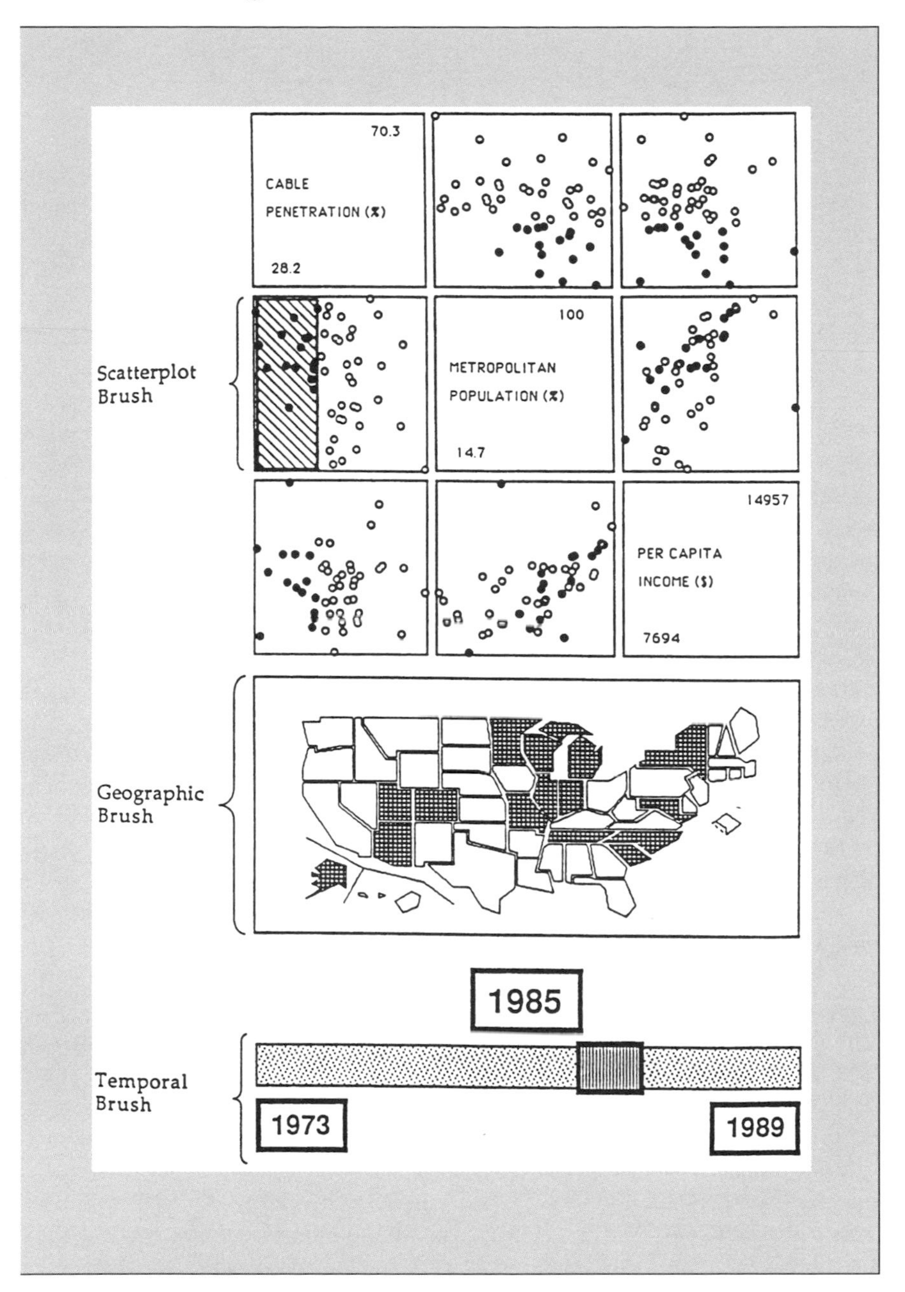
70.3
CABLE
PENETRATION (%)
28.2
100
METROPOLITAN
POPULATION (%)
14.7
14957
PER CAPITA
INCOME ($)
7694
Scatterplot
Brush
Geographic
Brush
1985
Temporal
Brush
1973
1989

location at different times. This violates a key assumption central to much statistical theory—that observations are mutually independent.

Attempts to deal with these challenges have stimulated the development of a new subfield of statistics. Although this work began in ecology and biostatistics and has lately attracted the interest of statisticians, many of these developments were pioneered by geographers. For example, geographers have developed methods for estimating the degree and nature of spatial autocovariance in point, line, and area data. They have also addressed such complicating features as periodicity or waves in spatial patterns. Geographers have also played an important role in developing multivariate statistical analysis methods to deal with the spatial and temporal autocovariance of much spatially referenced data.

Spatial data pose special problems that are subjects for research by geographers. Because geographic data often fail to meet distributional assumptions necessary for classical statistical procedures, geographers have been at the center of attempts to develop distribution-free methods for estimating statistical relationships among variables. They also have been involved in the development of methods for estimating prior probability distributions, either through Monte Carlo simulations that generate reference distributions unique to each locality or by developing Bayesian methods that allow investigators to incorporate knowledge of known relationships in statistical investigations.

Methods for evaluating data for spatial dependencies have received recent attention from geographers (e.g., Getis and Ord, 1992). Such measures are used both to identify spatial patterns in data and to allow analysts to understand spatial relationships in their data so that appropriate analytical techniques can be chosen. When measures of spatial dependency are applied to real world datasets of the magnitude needed to address the societal problems identified in Chapter 2, spatial dependency measures can become intractable. One promising solution to the dilemma is the application of massively parallel processing methods (Armstrong and Marciano, 1995).

Recent research in GISs aims to develop "data models" that facilitate the routine analysis of spatial dependence, spatial heterogeneity and spatially referenced diagnosis of regression models. These spatial data models are used to prepare data in the special forms needed to efficiently accomplish these spatial analysis methods. The intent is to bring the same level of enabling technology to spatial analysis that spreadsheets and statistical packages have brought to statistical analysis (Anselin et al., 1993).

Research in spatial statistics is believed by many to be in its infancy because many research questions are likely to yield to computational approaches. There are enormous complexities in the analysis of spatially referenced data that require additional research. For instance, research is needed to understand the scale dependence of statistical methods and sampling designs. This work is necessary to understand why different aggregations of spatially referenced observations give statistical results that vary dramatically and unpredictably from one spatial specification to another (e.g., Tobler, 1969). Research is also needed to understand how to separate environmental effects from other geographic variations in health research—how, for instance, to separate environmental factors that cause cancer from the density of susceptible populations when searching for cancer clusters. This last example illustrates the importance of research in addressing important public policy questions.

CONCLUSIONS

Current trends in geography's techniques suggest a future in which researchers, students, business people, and public policy makers will explore a world of shared spatial data from their desktops. They will request analyses from a rich menu of options, select the geographic area and spatial scale of analysis, and display their results in multimedia formats that are unanticipated today.

In developing GIS-GVis tools of the future, developers need to think more broadly about the context of the problems and the knowledge and skill levels of users. The users of tomorrow are likely to be a far broader group than at present—in background, perspectives, and skills. Indeed, many will be novices by today's standards, and they will rely on the embedded knowledge of experts in their use of these tools.

The users of the future will also bring different world views and theoretical perspectives to these tools. They will challenge the adequacy of current techniques for analyzing and understanding geographic phenomena, posing challenges that must be taken up by developers when designing next-generation tools and theories.

5

Geography's Contributions to Scientific Understanding

Geography contributes to science as a part of the broad, creative, multidisciplinary effort to advance the frontiers of knowledge. In so doing it offers significant insights into some of the major questions facing the sciences, related to the pursuit of knowledge both for its own sake and for the sake of improving society's well-being.

In this chapter, geography's contributions to scientific understanding, both actual and potential, are illustrated by way of example. The chapter itself is organized around the three "lenses" through which geographers view the world—integration in place, interdependencies between places, and interdependencies among scales. These constitute the major sections in this chapter, along with spatial representation. For each of these sections the chapter uses examples from geography's subject matter to illustrate how geographic thinking and approaches contribute to scientific understanding generally. The chapter then provides examples of how such thinking can be used to address important scientific and societal issues. These illustrations are followed by a similarly brief and selective discussion of spatial representation.

For the sake of brevity, and because the report is directed primarily to audiences outside of the discipline, the chapter illustrates geography's contributions by using a few examples, chosen primarily to illustrate the range of geographic research. Far more research by geographers has contributed significantly to science than is noted here specifically, of course. Examples of this research are referenced throughout this report.

INTEGRATION IN PLACE

Geography's traditional interest in integrating phenomena and processes in particular places has a new relevance in science today, in connection with the search for what some have called a "science of complexity."

Geography's Subject Matter

From its work on integration in place, geography has produced a substantial literature related to the challenges of integration in place and the significance of such integrative perspectives for scientific understanding. Two examples are *environmental-societal dynamics* and the *distinctiveness of place*.

Example: Environmental-Societal Dynamics

At least since Malthus,[1] the relationship between population and its social and environmental resource base has been a central issue for science, and geography has long focused on the nature of that relationship, ranging from local and contemporary contexts to global and historic processes. Geographers are involved in both data collection and analysis to identify connections among changes in population, environment, and social responses.

For example, geographers have reconstructed population-resource dynamics for a large number of places throughout the world. Following Butzer (1982), several important facets of the relationship over the long term may be distilled from these works, such as the following:

- Sustained population growth is not the norm at subglobal (regional) levels; given sufficient time, locales and regions may display significant declines in population.
- These declines are typically associated with political devolution.
- Populations do not always, or regularly, approach the limits allowed by the sociotechnological conditions in which they exist.
- Human-induced environmental change involves continual trial-and-error adjustments on the part of the population. Among agrarian-based societies, these adjustments are consciously related to a strategy to balance short- and long-term needs.

As another example, flows of materials, energy, and ideas across places have powerful impacts on human uses of the environment, and such impacts can mask basic understanding of contemporary environmental change. The sixteenth-

---

[1] The English economist and mathematician Thomas R. Malthus (1766–1835) is best known for his work on population and resources, embodied in his *Essay on the Principle of Population*, which was published in 1798.

century depopulation of the Americas is exemplary in this regard (see Sidebar 5.1). Spanish conquistadors reported large Indian populations throughout the Americas, but within 100 years these populations were reduced by two-thirds or more in Middle and South America as new diseases and sociopolitical orders were introduced from Europe and Africa. Not only is this event, which has received considerable research attention by geographers (Denevan, 1992), significant for historical demography and epidemiological history, it also has significant implications for understanding contemporary land cover change, a central issue in global change research.

Human impacts on the Earth have become sufficiently apparent and worrisome in the past few decades that the science of human-environment relationships has become a high-priority concern across disciplinary and national lines. Witness, for instance, the rise of the Intergovernmental Panel on Climate Change, the International Geosphere-Biosphere Programme, and the Human Dimensions of Global Environmental Change Programme. Geography is contributing signifi-

SIDEBAR 5.1 Pre-Hispanic Populations and Their Collapse

The collapse of pre-Hispanic native populations in the Americas is not seriously disputed, nor is its primary cause: introduced infectious diseases (Cook and Lovell, 1992). Yet some 400 years later the size of the Native American population during the contact period and the scale and trajectory of its collapse remain two of the most controversial issues in American demographic history. This controversy is perhaps most pronounced for the Basin of Mexico (now greater Mexico City). Eve-of-conquest population estimates for the basin range between 1 million and 3 million. Some 100 years later, only 70,000 to 350,000 Native Americans remained—a minimal loss of 65 percent of the original population.

Geographer Thomas Whitmore (1992) utilized a human-ecological simulation model to test various hypotheses for the size of the pre-Hispanic population and the trajectory of its collapse. The model accounted for demographic changes in age structure, mortality, fertility, and migration caused by introduced diseases, food shortages, homicide, and other factors; agricultural changes caused by excess mortality and ill health; and disease changes, that is, how individual epidemics affected human health and mortality. Exogenous forcing functions, such as extreme or poor weather, homicide, and labor withdrawal, were also addressed. The model was calibrated by using accepted or conservative estimates for parameter values and was fine tuned using sensitivity tests.

Three historical population reconstructions were simulated for the basin, ranging from mild to severe depopulation. The simulations supported "moderate" depopulation, in which a 1519 basin population of 1.6 million was reduced in a series of steplike catastrophes to about 180,000 in 1610—a 90 percent loss, with 80 percent of the loss occurring in the first 50 years. The simulations suggest that the most important factor in depopulation was short-term increases in mortality caused by disease. Other factors, such as homicide, were overwhelmed in importance by the virulence of new diseases in a population with no or little immunities.

cantly to agenda setting and research in these initiatives (Townshend, 1992; Henderson-Sellers, 1995; Turner et al., 1995).

Example: The Distinctiveness of Place

As noted in Chapter 3, one of the characteristic perspectives of geography is that place matters. In other words, *where* something takes place affects *what* takes place because of the mediating effects of local conditions. Geographers' concern with place leads them to explore not only the particular characteristics of individual places but also the processes by which humans divide up or appropriate portions of the Earth's surface for various purposes. As such, geographers direct attention to human *territoriality*, defined by Sack (1981, p. 55) as "the attempt to affect, influence, or control actions and interactions (of people, things, and relationships) by asserting and attempting to enforce control over a specific geographical area." Geographers are interested in human territoriality because the divisions of the Earth's surface reflect and shape the ways in which people think about the places where they live, as well as their decisions and actions.

It is impossible to understand human history fully, or past and present human actions, without reference to territoriality. Rather than simply asking questions about what happens in a given territorial unit, geographers consider how and why that unit came into being in the first place, and its history of development (Earle et al., 1996). Geographers have addressed a wide variety of issues related to territoriality, including disjunctions between political territories and environmental regions, ways in which territorial constructs affect ethnic relations, and the uses of territorial strategies to achieve social ends (Demko and Wood, 1994).

Geographic research on the distinctiveness of place addresses a range of economic, political, and social issues such as industrial agglomeration and regional economic development, the role of place-based political identities, places as foci for personal and societal opportunities or constraints, and the cultural meaning of the environment in which people live (see Sidebar 5.2). Research along these lines is shaping the way that place is conceptualized across the social sciences. For example, after a long history of asserting that geographic clusters of prosperity are temporary aberrations, economists now recognize that the evolving characteristics of places make such inequalities the rule rather than the exception (Arthur, 1988; Krugman, 1991). Geographers have gone beyond economic mechanisms to examine the role of political and social processes in constructing local "governance structures" that promote economic dynamism (Storper and Walker, 1989). This research attempts to make sense of evolving but persistent geographic inequalities in economic prosperity at all spatial scales.

Most studies of political preferences in modern states assume that differences are products of social cleavages along class, religious, or demographic lines. This assumption relegates place to a minor role in the political arena; to the extent that place-based identities and influences are considered at all, they are

SIDEBAR 5.2 The Rise of Silicon Valley

Geography's attention to place is exemplified by research on the extraordinary economic success of Silicon Valley. This success reflects the coming together of economic, political, and social processes in a *place*, creating new conditions that reinforce but may eventually undermine economic success (Hall and Markusen, 1985; Scott, 1988a; Saxenian, 1994).

Silicon Valley had appropriate preconditions for high-technology industry, showing the importance of certain local conditions as a trigger for agglomeration. Yet at the same time these conditions did not predetermine the success of Silicon Valley because they could be found in a number of other places. Geographers have demonstrated how the growth of industrial districts in new locations such as Silicon Valley and southern California reflects the rise of new industrial sectors with new economic and labor requirements and, in some cases, a desire to develop their activities at some distance from locations associated with more traditional modes of production (Storper and Walker, 1989). A combination of competition and cooperation accelerated technological change and secured market niches for selected firms.

In places like Silicon Valley, there is a mutually reinforcing feedback between place characteristics and economic activities. This feedback reflects economic interdependencies and political interventions in supporting economic growth and demonstrates how agglomeration generates increasing returns, which create economic dynamism. Silicon Valley is an example of a new and distinctive combination of economic, political, and social activities, now developing in just a few places, with broader ramifications for many other locations (Scott, 1988b, c).

Certain conditions in Silicon Valley may eventually lead to an undermining of economic growth. These conditions include high labor costs, which are inducing the relocation of low-wage, less skilled employment to Southeast Asia and some skilled software development work to India and China. These conditions also include congestion, environmental pollution, and even poverty and reinforced social inequality between higher-paid salaried employees and lower-skilled wage workers (Saxenian, 1994). In addition, the decentralized organizational structure of Silicon Valley may also hinder longer-term economic prosperity because it encourages excessive competition, hypermobility of skilled labor, and industrial fragmentation (Florida and Kenney, 1990).

The questions of how places adjust their local conditions to retain economic dynamism and who in those places benefits from this dynamism are issues of continuing research, for both those concerned with the geographic organization of economic activities within the United States and those concerned with reinforcing the competitive edge of the United States in the world economy.

treated as anachronisms that have resisted the general trend toward the nationalization of political life. That is, where people live is thought to be of minor importance. Geographic research, however, has established the importance of "the fluid, constantly reworked local political cultures of particular places" (Agnew, 1992, p. 68), demonstrating the continued importance of the experience of place in the political process.

Similarly, geographic work has highlighted the importance of place in the formation of cultural and social identities and experiences. In this late-twentieth-century world, as Americans struggle to address the tensions and celebrate the richness of human differences—of ethnicity, race, nationality, gender, and generation—a focus on the ways that ideas about place serve to divide people but also to connect them can offer new visions for personal and social values (Agnew, 1987).

Geographers have also challenged the tendency of much social science research to treat the environment in which people live merely as a passive byproduct of history. They have argued that the material characteristics of the environments in which people live reflect and influence personal, social, and environmental understandings. As the social sciences begin to take more seriously the role of symbols and images in human affairs, geography's concern with the social dimensions of landscapes has taken on new relevance and visibility. Geographers have done a great deal of research to uncover the political/social meanings, influences, and conflicts embedded in landscapes (e.g., see Cosgrove and Daniels, 1988; Anderson and Gale, 1992) and representations of landscapes (e.g., Harley, 1990; Pickles, 1995b).

By focusing on the tangible environments where people live and work, geographic research is part of a growing thrust within the social sciences to understand the importance of everyday life in social change (e.g., Giddens, 1985). At the same time, geographers are affecting the direction of that thrust by linking human ideas and actions to the settings in which they are embedded.

Geography's Relevance to Issues for Science and Society

Geographic research addressing integration in place has put the discipline at the frontier of experimentation with integration as a challenge for science. Geography's experience with integration in place also has been fruitful in providing insights to issues of interest to science at large, as illustrated by the following examples of *complexity and nonlinearity* and *central tendency and variation*. Geographic research on integration in place is also important to scientific understanding of important societal issues. Three examples are given below—on *economic health*, *ecosystem change*, and *conflict and cooperation*—to illustrate this importance.

Example: Complexity and Nonlinearity

Places are natural laboratories for the study of complexity because places exhibit a wide array of interlocking processes and activities, as well as interconnections with other places. Nonlinear growth and decline also are found locally in part because introduced processes or activities may not encounter well-developed moderating influences. Geographers have examined the complex and nonlinear systems of places to better understand how and why places change. Historical research on the evolution of the American urban system, for example, illustrates the evolutionary nature of human settlement systems. This research shows that early settlement patterns can create "path dependencies" for the future evolution of settlement systems. It also shows how economic restructuring, such as the shift from mercantilism to industrial capitalism, can create "bifurcation" of settlement systems with new nodes of growth in some regions and dissipation of growth in others (Borchert, 1967, 1987; Conzen, 1975; Dunn, 1980; Pred, 1981). Theoretical research has identified how the complexity of spatial economic dynamics reflects disequilibrating contradictions and social conflicts, resulting in periodic attempts by the private sector and the state to overcome emerging conflicts and crises through spatial restructuring (Harvey, 1982; Sheppard and Barnes, 1990).

Geographers have applied systems theory to help understand the complex interactions between nature and society that are caused by natural hazards, including multiple adjustments and attendant feedbacks (Cutter, 1993). Geographers have also examined the mechanisms of ecosystem stability and change, especially human and other agents of short- and long-term ecosystem change (Zimmerer, 1994). Ideas about chaotic behavior or catastrophic events within places, additionally, have contributed to research on growth within and among cities (Allen and Sanglier, 1979; Dendrinos, 1992). These studies illustrate geographers' contributions to a more fundamental understanding of environmental and social systems in ways that should engage ecologists, engineers, mathematicians, physicists, and other members of the scientific community.

Other geographic research has been directed toward the identification and description of patterns that may have emerged from nonlinear, complex, or chaotic dynamics. Fractal dimensions, in particular, have been used to simplify and represent the outcomes of nonlinear, chaotic, or complex dynamics (Goodchild and Mark, 1987). Urban settings (Batty and Longley, 1994) as well as satellite and map images (Malanson et al., 1990) have been usefully analyzed and characterized with fractals.

Example: Central Tendency and Variation

Interactions in space and with nature tend to result in certain spatial and environmental regularities, leading to the study of expected outcomes, or central tendencies, across geography's domains of interest (Chorley and Haggett, 1967). Geographers have recognized, however, that observable geometries in the social and physical worlds are dynamic in their nature and multidimensional in their explanation. Certain geographic patterns reflect efficiency (as in economic production systems), but only under rather narrowly defined conditions that are subject to change (such as the time or cost of travel) and to inherent variability. Together, change and inherent variability often influence the observed variation, which can take the form of unsystematic departures from central tendency, changes to the central tendency itself, or alterations in the variance structure. Changes in variation can signal shifts from one system state to another; therefore, variation cannot be ignored without deleterious or sometimes catastrophic consequences. Geographic research into the nature of change and variability, as well as into central tendency, has revealed much about the dynamics of places (Dendrinos, 1992). As in other sciences, geographers also have recognized that variation and central tendency are usually interdependent and cannot be evaluated or understood separately.

Example: Economic and Social Health

A geographic perspective recognizes that economic changes can create or exacerbate economic imbalances across places, whether or not the economic system overall is trending toward or away from equilibrium. A particular concern of geographers is the implications of economic change for different groups in society within a place, especially for groups distinguished by class, gender, and race. Related issues include the composition of the work force as rooted in social forces and potentials for cooperation versus conflict (see Sidebar 5.3).

Geographers have examined high-technology centers to evaluate their potential as models for regional growth in other areas (see Sidebar 5.2). They have noted that locational considerations are different for innovation centers than for other industrial activities such as branch plants (e.g., high skill levels are especially important for innovation centers). Since labor is less mobile than capital, regional growth related to technological change is likely to follow existing patterns of labor skills, which increases the challenges for areas that do not now have competitive skill levels (Malecki, 1991).

Example: Environmental Change

Scientific concerns about environmental change have increased markedly in the past few decades. Geographers have made important contributions to the understanding of such changes through their research on human-induced climate change, ecosystem dynamics and biodiversity, and earth surface processes.

For example, human populations are increasingly concentrated in urban and suburban regions. Land surfaces in these areas, in turn, are being transformed into highly unnatural mosaics, mosaics often dominated by interconnected and impervious patches of buildings and transportation networks. With the transformation of rural landscapes into suburban and urban landscapes comes dramatic changes in local and regional climates (see Sidebar 5.4). Urban heating and drying, for example, have been measured and simulated by geographers for decades (e.g., Terjung and O'Rourke, 1980; Arnfield, 1982; Grimmond and Oke, 1995). Geographers' research not only has brought to light climatic consequences of urbanization, but their models have begun to provide a means for assessing the potential climatic impacts of future urbanization.

Another focus of environmental change research involves reconstructing recent disturbance patterns and ecosystem processes in forest, shrubland, and desert communities through careful fieldwork and historical analyses (see Sidebar

SIDEBAR 5.3 Civil Unrest in Los Angeles

On April 29–30, 1992, thousands of Los Angeles residents engaged in looting, arson, and other violence within concentrated areas in the city's south-central section, following the announcement of a not-guilty verdict in the jury trial of city police officers accused of using excessive force in the arrest of an African American motorist. Quelling the violence and fighting fires required 22,700 police officers, firefighters, National Guard troops, and other U.S. military personnel. More than 16,000 people were arrested, more than 2,300 were injured, and 43 people died. Property damage was estimated at $750 million to $1 billion.

Work by Johnson and colleagues (Johnson et al., 1992) suggests that this civil unrest was the result of a confluence of several economic, social, and political changes—some international, some national, and some local—that set the stage for a social "explosion," given the appropriate provocation. Local changes in the economic base, related to the high level of plant closings in central Los Angeles and increased employer options within the larger metropolitan area (see Figure 5.1), combined with international migration, caused dramatic changes in the area's cultural, ethnic, and socioeconomic composition (see Figure 5.2). These changes heightened local social instability and tension.

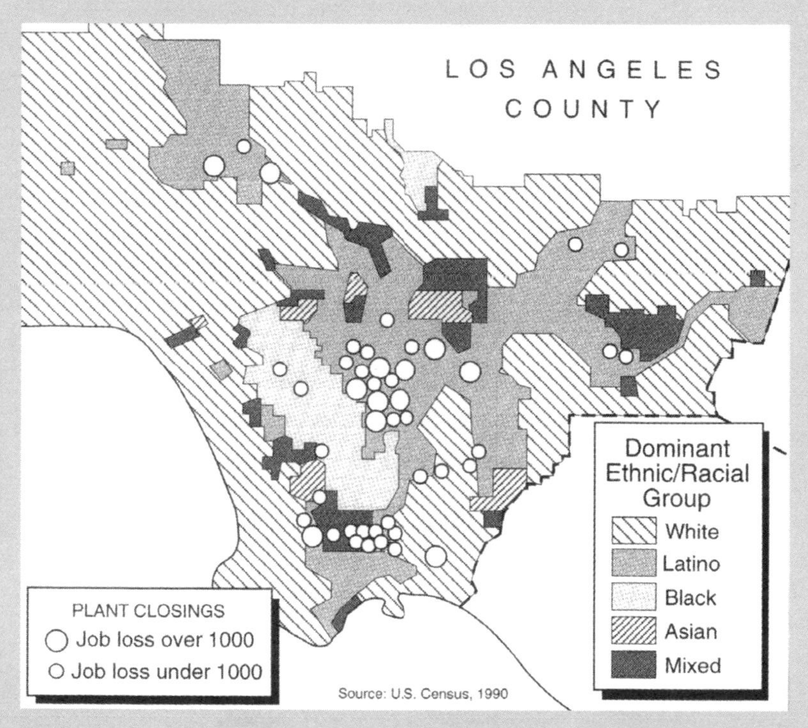

FIGURE 5.1 Plant closings in Los Angeles County from 1978 to 1982. Source: Johnson et al. (1992).

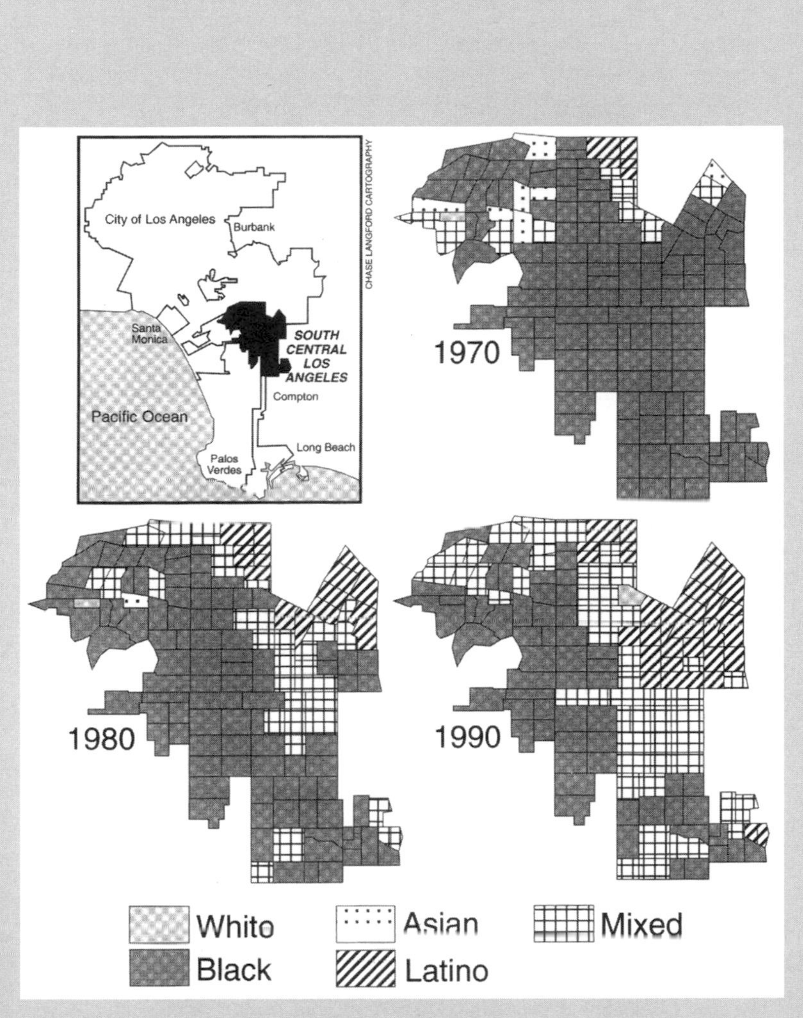

FIGURE 5.2 Ethnic change in south-central Los Angeles, 1970–1990. Source: Johnson et al. (1992).

SIDEBAR 5.4 Urban Climatology

Urbanization dramatically alters the land surface and converts preurban local or regional climates into distinctively "urban climates." Probably the best known and most intensively studied urban-climatic feature is the "urban heat island," although considerable attention has also been directed toward the effects of urbanization on precipitation, humidity, wind, and the air quality regimes of cities. Using integrated programs of fieldwork and numerical modeling, geographers have been at the forefront of assessing urban influences (especially the effects of urban surface materials and morphology) on local and regional climates (Oke, 1987). Research by geographers also is beginning to suggest that extensive land surface changes associated with urban- and suburbanization occurring worldwide may be contributing to global climate change.

Sue Grimmond and Tim Oke have been particularly effective at integrating in-the-field measurement programs with the numerical modeling of urban climates (Grimmond and Oke, 1995; Grimmond et al., 1996). Making and evaluating heat and moisture flux observations, as well as compiling surface character databases, their research teams have examined a number of North American cities, including Los Angeles, Chicago, Miami, Vancouver, Sacramento, Tucson, and Mexico City. Not only have they documented the considerable variability that exists both within and between cities, but their analyses show that daily patterns of the fluxes and the timing of the peaks are remarkably similar among the cities. Their measurements further indicate that evapotranspiration is even higher than expected in many residential areas, owing to the irrigation of planted vegetation. Evapotranspiration in other parts of the city tends to be quite low, as available energy mostly warms the urban fabric.

Grimmond and colleagues have also been able to use geographic information systems (GISs) to help synthesize land surface information, field measurements, and model simulations (Grimmond and Souch, 1994). Their innovative approaches are revealing the often elusive source regions of the heat and moisture fluxes (e.g., evapotranspiration), as well as the character of the land cover in those source regions (see Plate 6). Although others have investigated source regions, Grimmond and colleagues are identifying and quantifying them more precisely than ever before and in turn are clarifying the spatial and temporal relationships between urbanization and attendant climate change. Their results have the potential to help isolate the influences of built environments on global climate change.

5.5). Geographers also are addressing ecosystem disturbance and change over longer time scales through analysis of lake sediments from a variety of ecosystems (Horn, 1993; Liu and Fearn, 1993; Whitlock, 1993). The pooling of paleoenvironmental datasets over large regions has allowed geographers to map species ranges and ecosystem boundaries for selected times during the past 2 million years (Wright et al., 1993). These maps document the biotic response to past global changes and also provide a means of evaluating models of the Earth's climate system.

One of the most pressing issues for global and regional environmental change is ecosystem change, including the loss of biodiversity (USGCRP, 1994). Geography has a long tradition of studying landscapes, particularly the impacts of physical and human processes on landscapes and their ecosystems. For example, geographers study the distributions of plant and animal species and how these distributions are shaped by local and regional environmental conditions—including human activity—and by human-influenced migration and selection (Sauer, 1988). Geography also has a long tradition of studying the spatial patterning and

SIDEBAR 5.5 Disturbance Regimes Along Environmental Gradients

Research by geographer Tom Veblen and colleagues (Veblen and Lorenz, 1988; Veblen et al., 1989, 1992) on the effects of fire on ecosystems of southernmost South America is helping science to understand the ecological effects of natural and anthropogenic disturbances. Veblen and colleagues have reconstructed the spatial and temporal variations in fire frequency and size along a transect from temperate rainforest to steppe in northern Patagonia (see Figure 5.3). They collected data on fires in the region by using tree-ring data for the period 1722–1991 and national park records for 1938–1989 (Kitzberger and others, unpublished). Fire history data were collected for several vegetation types and precipitation zones so that climatic and human influences could be determined.

Over the entire period examined (1722–1991), there was a general spatial trend of increasing fire frequency from the western rainforests to the moderately dry (but still forested) central sector, followed by a decrease at the limit of tree growth at the steppe. Maximum fire frequencies in the intermediate portions of the transect coincided with an optimal combination of adequate moisture availability for the development of forest cover and sufficiently dry springs and summers for fuel desiccation.

Human impacts on fire regimes were highly differentiated according to position along the transect. In the dry vegetation types there was a substantial increase in fire frequency in the mid-1800s associated with the immigration of Indians from Chile. Aboriginal occupation of the mesic forests was sparse, in contrast, and here fires are infrequent until white settlement in the 1890s. The most dramatic increases in fire frequencies are within the mesic forests that European colonists burned extensively in failed attempts to establish cattle pastures. After the 1920s, fire frequencies declined in both wet and dry vegetation types, reflecting the initiation of fire suppression throughout the transect and, in the steppes, the cessation of Indian-set fires for hunting.

Climatic impacts on fire regimes also varied with position along the transect (Kitzberger and others, unpublished). Fire in wet forests is strongly favored by relatively short droughts during spring and summer; even short dry periods are sufficient to dry out the bamboo understory and provide fuel. In dry vegetation types, fires correlate not with short-term droughts but with one- to two-year periods of drier-than-average conditions—especially after wet periods that enhance fuel production.

This research demonstrates the need to consider natural and human disturbances in tandem, rather than in isolation, and to examine spatial variations in disturbances to explain vegetation patterns at the landscape scale.

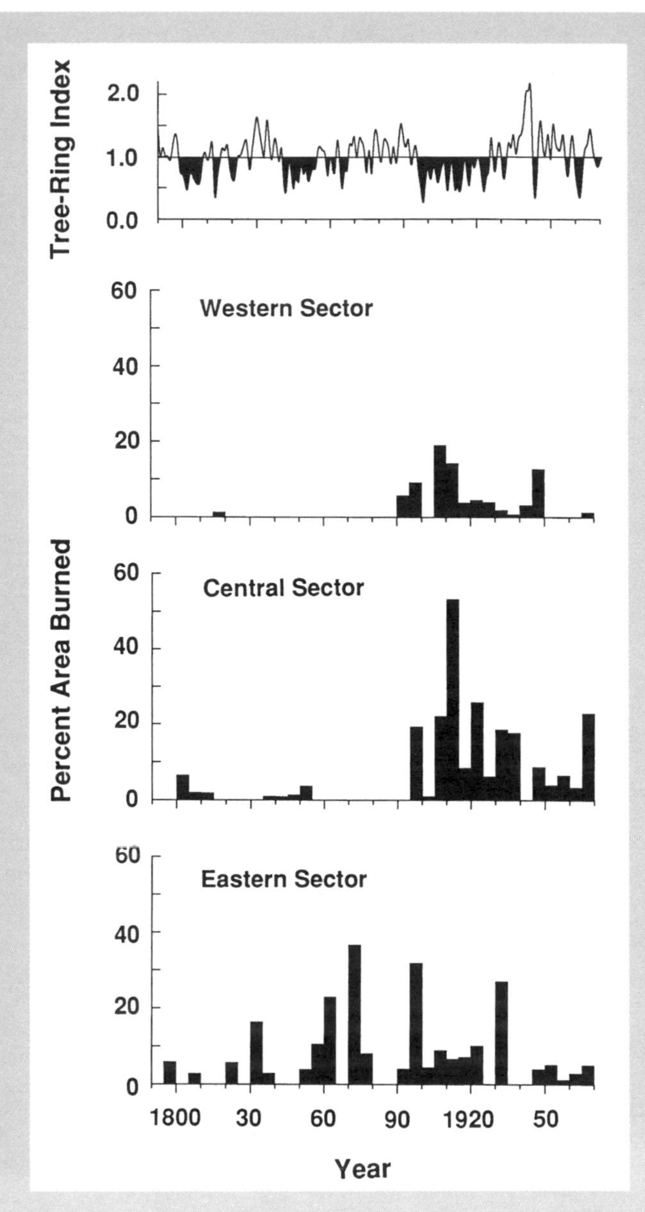

FIGURE 5.3 Percentage of areas burned by five-year periods along a transect from Andean rainforests (western sector) through mesic forests (central sector) to xeric woodlands (eastern sector) and a drought-sensitive tree-ring chronology reflecting regional climatic variations in northern Patagonia. Source: Veblen et al. (1992).

human and nonhuman determinants of biodiversity within both "natural" and agricultural landscapes. This tradition predates recent concerns about biodiversity loss.

An important focus of recent geographic work has been spatial variations in the nature, recurrence, and biotic consequences of human and natural disturbances such as fires, treefalls, forest clearance, and floods (Vale, 1982). This research provides essential knowledge for devising systems to preserve biodiversity at local, regional, and global scales (Baker, 1989a; Young, 1992; Medley, 1993; Savage, 1993).

In their focus on Earth surface processes, geographers are paying increased attention to the nature of change itself, and the transitions between different change states. There also is increasing interest in the flows of energy and mass through and across the Earth surface system as an avenue to understanding the underlying structure of environmental change. Geographic investigations explore such changes on time scales ranging from less than a single year to hundreds of thousands of years.

At time scales of decades to centuries, geographic work is concerned mainly with documenting changes in Earth surface systems and assessing underlying causes. One focus of geographic research, for example, involves the reconstruction of historical dimensions of glaciers through photographs and surveys in order to assess regional climate change (e.g., Chambers et al., 1991). Another important focus of geographic work at these time scales concerns the effects of human settlement on river systems—for example, the work of Kesel et al. (1992) on the effects of human settlement on the sediment load of the Mississippi River; several decades of work by M.G. Wolman and his students on the effects of urbanization on water and sediment runoff to rivers; work beginning with Grove Karl Gilbert on the impacts of mining on river systems (see also James, 1989; Mossa and Autin, 1996); work by T. Dunne and other geographers (e.g., Abrahams et al., 1995) on land-use changes in developing countries on slope and stream processes (see Sidebar 5.6); and work by Trimble et al. (1987) on the effects of revegetation on river dynamics.

At time scales of 10,000 to 100,000 years, another concern of geographic research has been understanding connections between climatic changes and the Earth's physical response, such as the effects of orbital changes on the amount of effective solar radiation received at the Earth's surface (Cervany, 1991). These so-called Orbital or Milankovitch changes have been used to explain periodic "floods" of icebergs in the northern Atlantic and other ocean surface responses (Broecker, 1994), and terrestrial changes in climate as recorded by rock varnish (Liu and Dorn, 1996). Evidence for terrestrial climate change has been documented from such diverse sources as wind-deposited silts (loess) on the Great Plains (Feng et al., 1994), lake fluctuations in the Great Basin (Currey, 1994), and glacial moraines in the Sierra Nevada (Scuderi, 1987).

SIDEBAR 5.6 Land Use and Soil Erosion

Soil erosion and sediment deposition are essential landscape-shaping processes that become particularly important to human occupants of a place when they reduce the productivity of agricultural lands or decrease the capacity of reservoirs built for hydroelectric power generation. Geographer Carol Harden has studied soil erosion in inhabited watersheds of North, Central, and South America: how and where it occurs, how it affects and is affected by farming families, and how the eroded soil moves through the landscape to be carried away by rivers or trapped in reservoirs. Harden uses a portable rainfall simulator to repeatedly produce small (15-cm-diameter) "rainstorms" at many different locations. During each 30-min "rain" experiment, she measures the amount and rate of the "rain," observes whether water is absorbed by the soil or becomes runoff, and collects and measures the runoff and eroded sediment. Repeating these experiments across different soils and land uses, she has demonstrated that soil erosion is more spatially complex than often portrayed and that basin-wide soil loss is strongly impacted by two classes of land use that are routinely ignored in watershed models: "abandoned" land and roads and trails.

In a detailed study in the 5,186-km^2 Paute watershed in highland Ecuador, Harden (1991, 1996) found that abandoned or fallow fields had significantly higher runoff coefficients than either cultivated fields with mature crops or bare fields that had recently been plowed (see Figure 5.4a). Sediment detachment was also high (Figure 5.4b), while soil carbon (reflecting organic content) was very low (Figure 5.4c). Harden attributes these patterns to the fact that, while these abandoned/fallow lands in highland Ecuador are abandoned from a management point of view, they are still used by local people, who view them as common property. Unregulated grazing of domestic animals reduces vegetation cover and compacts the soil, resulting in high runoff coefficients.

Moisture stress resulting from low precipitation and rapid runoff further impedes vegetation recovery, leading to accelerated erosion and a devastating cycle of positive feedback that further degrades both soil and vegetation. Harden's work showed that abandoned/fallow lands, which are often regarded as having no land use, are key to understanding and modeling soil erosion in inhabited landscapes and must be included in studies relating land use to soil erosion.

Further rainfall simulation experiments, comparing paired road (path) to off-trail sites in Ecuador, Costa Rica, and the United States, revealed that roads and footpaths are a second category of land use whose erosional contributions are extremely important but inadequately assessed by traditional modeling approaches, which weight land cover types by their area. Harden's research demonstrates that roads and footpaths are the most active runoff-generating components of many tropical and temperate landscapes (Harden, 1992; Wallin and Harden, 1996), producing runoff earlier than other surfaces in the same rain event, generating runoff under rainfall that is too light to yield runoff elsewhere and effectively extending the drainage network. Although roads and footpaths occupy only a very small proportion ($<1\%$) of the Earth's surface, Harden's work argues that they profoundly affect geomorphic/erosional processes and therefore must be incorporated into hydrologic and soil erosion models.

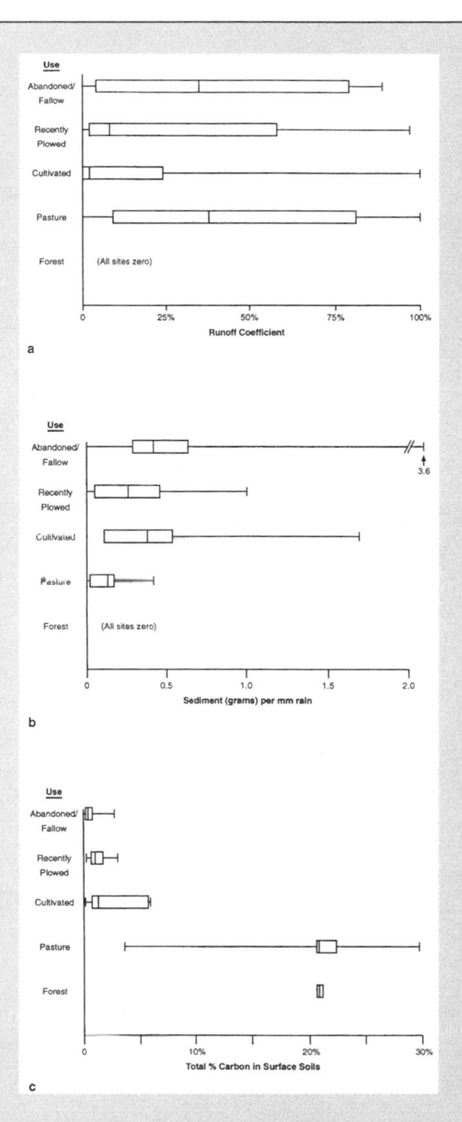

FIGURE 5.4 (a) This box plot of runoff coefficients (runoff/rain-fall) from simulated rainfall on prewetted soils demonstrates the variability of runoff and the high level of runoff production on abandoned fields and pastures. (b) The impact of land abandonment on soil erosion becomes evident when high runoff, as shown in part (a), is coupled with high rates of sediment detachment. (c) Very low percentages of carbon in soil samples from abandoned fields attest to the ongoing cycle of soil degradation that occurs where vegetative recovery is impeded.

Example: Conflict and Cooperation

In any effort to understand how individuals and groups relate to one another, context is fundamental. Geography's concerns with the integration of phenomena in place and the positioning of one place with respect to others are key to an understanding of context; they focus attention on the importance of such matters as resources, land use, and the distribution and movements of peoples. Geographic work highlights the connections between social forces and the material and spatial circumstances in which they are embedded.

For example, geographers have shown how conflicts over water have affected everything from territorial disputes in the Middle East (Kliot, 1994) to gender relations in West Africa (Carney, 1993; Schroeder, 1993). Research into the so-called urban underclass has shown how the geographic concentration of minority populations is contributing to their alienation as a result of discrimination and suburban exclusion in the housing market, suburbanization of well-paid jobs, inadequate financing of education in central cities, and the out-migration of better-off ethnic minorities who have gained access to suburban housing markets (e.g., Jackson, 1987).

Studies along these lines contribute to larger efforts to understand the nature of social and ethnic conflicts. They point to the necessity of moving beyond sociological analysis to understand how the material and spatial attributes of specific places affect the formation and interaction of social and ethnic groups. Such studies provide insights into connections and relationships that matter in the ongoing interdisciplinary effort to better understand the forces shaping conflict and cooperation.

INTERDEPENDENCIES BETWEEN PLACES

In many ways, geography is a science of flows. It sees the world not as a static mosaic of spatial units but as an ever-changing tapestry of landscapes, movements, and interactions. As noted in Chapter 3, geographers recognize that "place" is defined in part by the movement of peoples, goods, and ideas from other locations.

Geography's Subject Matter

Studies of interdependencies between places are well represented in geography's literatures. For example, for more than a generation, geography has been a leader in improving quantitative models that help to explain, predict, and optimize spatial interactions. Contemporary work in this field seeks to incorporate behavioral dimensions of spatial interaction and to capitalize on advances in spatial econometrics. Although there has been heated debate over the meaning of mathematical formulations of such models, their continued widespread application is testimony to their usefulness in many practical situations.

Contributions by geographers to our understanding of the interdependencies between places are illustrated by studies of *spatial economic flows*, *human migration*, and *watershed dynamics,* as illustrated in the following subsections.

Example: Spatial Economic Flows

Following the basic work of Wilson (1974) and others, geographers have researched the movement of people, commodities, and capital and the spatial choice patterns of consumers in relation to alternative service sites. This research addresses spatial interactions of individuals at the microlevel and interregional flows at the macrolevel.

At the microlevel, geographers have observed that patterns of spatial interaction differ by socioeconomic class and gender (Hanson, 1986), affected by such characteristics as income, family responsibilities and geographic relationships within an extended family, and the experience and expectations of the individual and of those with whom the individual interacts. To the extent that these effects can be modeled and generalized, they help geographers understand the operation of local labor markets, shopping patterns, and information diffusion. An interdisciplinary body of research by geographers, economists, and sociologists has indeed shown that one of the most persistent empirical correlates of commodity and population flows is distance, even in situations where standard economic and sociological variables perform inconsistently. Geographers argue, however, that distance itself is not a datum but a social construction whose influence changes with shifts in the barriers between, and communication technologies linking, different places.

Data on interactions among places (e.g., population migration, technological diffusion, and commercial trade) are less commonly available than data on analogous characteristics in individual places. The problem is compounded by the multiple geographic scales at which interactions occur. For example, case study, survey-based data suggest that trade between states in the United States has probably been increasing over the past two decades, yet more is known about each state's international trade than about its trade with other parts of the country. Figure 5.5 shows purchases and sales of selected Washington State firms with other regions of the United States. Although spatial patterns of sales and purchases vary significantly from state to state, there is a symmetry of states' sales to and purchases from interstate regions, despite the different nature of goods imported and exported from a given state. Correlation coefficients for state firms' sales and purchases by region are on the order of 0.7 and are highly significant (Beyers, 1983).

Modeling of spatial interaction data is at the heart of geographic analysis. The expansion method of generating models that embed temporal or spatial shifts

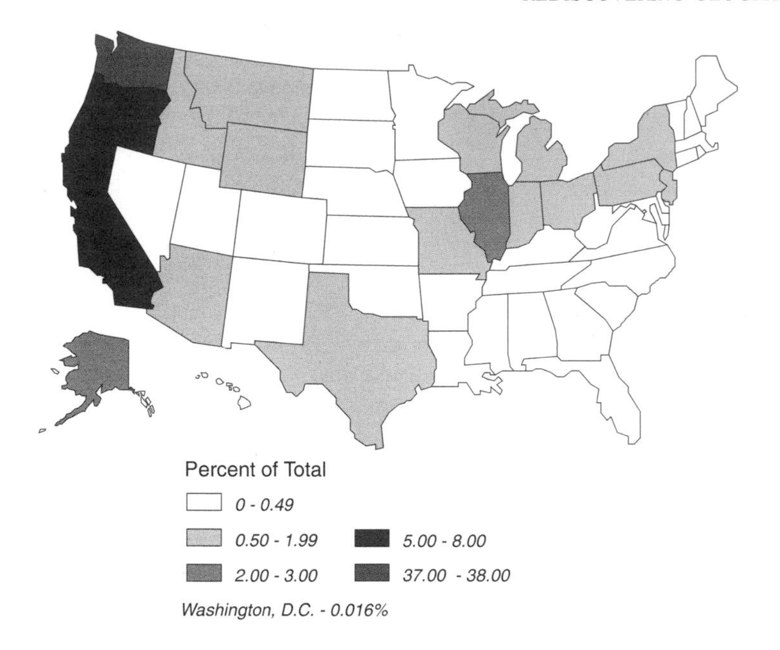

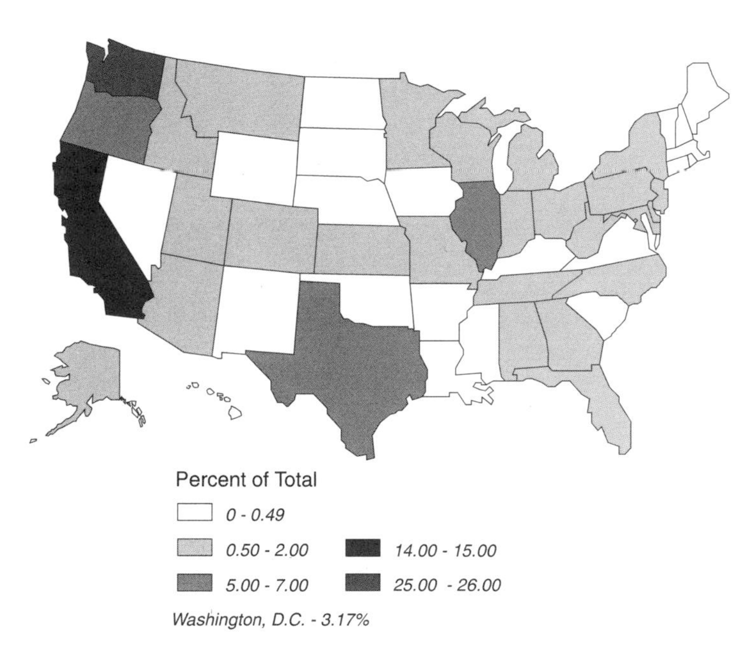

FIGURE 5.5 Purchases and sales of a sample of Washington State firms from (top) and to (bottom) other regions of the United States. Source: After Beyers (1983).

in key parameters allow researchers to uncover greater specificity of spatial relationships. This method has gained widespread use in geographic analysis, from its introduction into the literature (Casetti, 1972) to its use in a range of applications and interpretations (Jones and Casetti, 1992). Research by geographers and regional scientists has shown how to derive spatial interaction models based on either traditional information theory or optimal decision-making theory. This theoretical work has been extended to analysis of the interaction between consumers and suppliers of services. Spatial interaction simulations can pose "what if" questions about retail patterns and behavior similar to the questions about the flows of goods among states. Much of the literature in relatively new academic journals such as *Geographical Systems*; *Location Science*; and *Computers, Environments, and Urban Systems* contains illustrations of such models.

Example: Human Migration

Decisions to relocate are among the more important decisions made by households, with far-reaching implications for the links between places. Conceptualization of the search and selection process by Wolpert (1965) and Brown and Moore (1971) has been formalized in a model of decision making and housing search under uncertainty (Smith et al., 1979). This model incorporates both preferences and expectations of relocation decisions and provides important insights into household searches within the residential environment.

Recent work on modeling of migration and mobility seeks to address the dynamic nature of the process and the way in which decisions to move are related to age, family composition, and economic circumstances (Clark, 1992; Clark et al., 1994). For example, one of the strongest microlevel determinants of whether individuals are likely to move is age or stage in the life cycle (see Sidebar 5.7). During the 1970s, all of these influences were evidenced as the extremely large baby boom cohort (people born from 1946 to 1964) passed through the peak mobility ages (ages 20–34).

Few social science variables can be confidently forecasted far into the future. Barring major calamities, however, the inexorability of the aging process makes future age composition one of the best independent variables for population forecasting applications. As geographers learn more about these demographic influences on migration, population analysts should become better able to inform public policy at both national and local scales.

Example: Watershed Dynamics

Through their research, physical geographers have demonstrated the importance of interdependencies between places on understanding the environment. A major contribution to research on river ecosystems, for instance, has been the

SIDEBAR 5.7 Impacts on Interregional Population Movements

Recent geographic research on shifts in regional age composition has exposed some of the underlying reasons for changes in U.S. mobility and broad-scale patterns of interregional population movement, such as the large frostbelt to sunbelt streams of net population exchange during the 1970s. Using ideas similar to Easterlin's (e.g., Easterlin, 1980) on how the sizes of successive generations influence levels of fertility in a society, Plane and Rogerson (1991) investigated the age-specific geographic mobility of young adults ages 20 to 24 from 1949 to 1987 as a function of that age group's overall share of the U.S. population (see Figure 5.6). They observed an inverse relationship between mobility and cohort

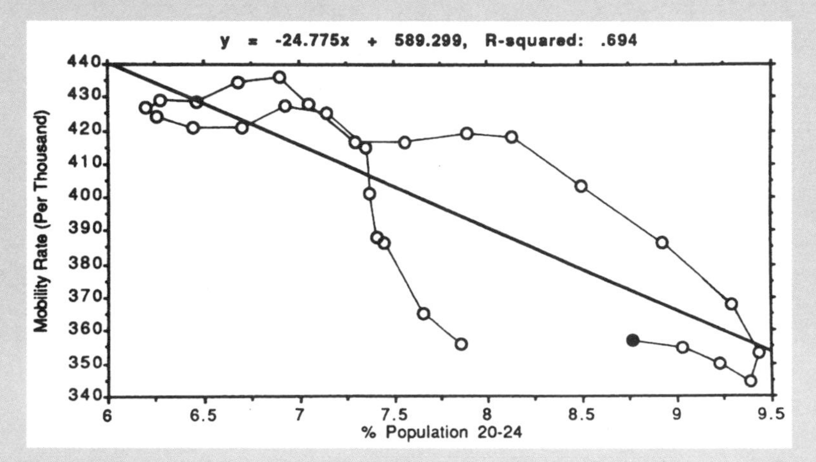

FIGURE 5.6 The mobility rate of young adults ages 20 to 24 as a function of that group's overall share of the U.S. population. The data points on the figure represent three-year moving averages. The filled circle is the average for the period 1985–1987, the latest data available at the time the research was completed. There are gaps in the data between 1971 and 1975 and 1976 to 1980. Source: Plane and Rogerson (1991).

recognition and analysis by geographers of spatial connections and long-distance impacts. Although spatial analysis of river behavior began in the 1940s, a thorough understanding of the geography of processes has emerged only recently. Until the mid-1970s, many natural science disciplines addressed the workings of individual ecosystem components and their connections to adjacent components. The description and analysis of riparian habitats critical to desirable or endangered species, for example, involved a focus on local dynamics of vegetation, soil, and water. Similarly, the behavior of rivers was understood in terms of the hydraulics and mechanics of the materials at a given location. This focus on analytic approaches improved scientific understanding of local processes, but it was less successful in predicting externally induced changes in these environments.

size, which they suggest is the result of the different economic and sociological fates that await individuals from generations of different sizes as they age in an economy that favors economic growth through what is known as "spread effects." In labor and housing markets, stiffer competition faces individuals from large cohorts, making them less likely to move. When there are large numbers of labor force entrants, wage levels are pushed down and fewer good jobs are available, thus reducing economic incentives to move. When many people first enter housing markets, additional units must be supplied and prices rise.

The relationship shown in Figure 5.6 is strikingly similar to the boom-and-bust cycle of mobility discussed by Plane (1993; see Figure 5.7). In this cycle, mobility lags slightly behind cohort size during successive generations, tracing a "figure eight" path on the graph.

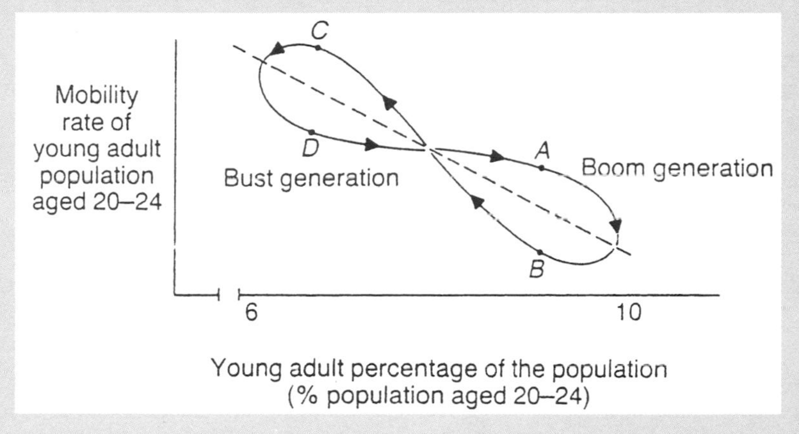

FIGURE 5.7 Expected mobility rate of the young adult population (ages 20–24) as a function of that cohort's size relative to the rest of the population. Boom cycles describe periods when there is a large increase in young people; bust cycles describe periods when there are decreasing numbers of young people. Source: Plane (1993).

Beginning in the mid-1970s, physical geographers (and scientists in other disciplines who use geographic perspectives) adopted a more holistic view that emphasized spatial patterns, connections, and long-distance impacts. Riparian habitats were seen to respond to changes in the watershed upstream, as well as to local dynamics. For example, William Baker (1989b), Jacob Bendix (1994), and George Malanson (1993) have shown that the composition and dynamics of riparian forests depend both on local conditions and on the location of forests in the stream network and the distant areas that contribute water and nutrients. Similarly, analysis of the movement of pollutants in watersheds to the Chesapeake Bay and other East Coast estuaries showed how our understanding of estuarine environmental quality could be enhanced through analysis of events in the upstream watersheds (Marcus and Kearney, 1991). Using theoretical constructs from this work, the U.S. Environmental Protection Agency (EPA) has developed more effective monitoring and remedial measures to control contamination from runoff into these estuaries.

Geomorphology has also become more concerned with the spatial perspective, and geomorphic systems analysis has been expanded to incorporate location and spatial connections for measuring and mapping physical forces and stresses, hydraulic resistance, and sediment yields. The result has been greater effectiveness in predicting environmental changes at critical locations—for instance, at a salmon spawning area in a river—based on system-wide changes that are connected in space by slopes and channel networks. As a consequence, geomorphology became more useful to society: geomorphologists now participate in U.S. Department of Agriculture field units, EPA investigations of bridge sites and other civil works, planning for geomorphic hazard mitigation, evaluations of critical habitats, and efforts to stabilize public lands.

Relevance to Issues for Science and Society

Spatial interdependence is an issue of great importance in a wide range of sciences, from physics and astronomy to climatology and geopolitics. Geography's perspectives on this phenomenon have contributed to our understanding of several issues of interest to science generally, including *complexity and nonlinearity* and relationships between *form and function*, as illustrated by the following examples. Geography's concern with spatial interdependence is also directly relevant to the base of scientific knowledge related to critical issues for society—as shown by subsequent examples on *conflict and cooperation* and *human health*.

Example: Complexity and Nonlinearity

A distinctive contribution of geographic research to the theory and modeling of complex systems (Pines, 1986) is the recognition that changing patterns of interactions between places can be a significant source of complexity. This observation has been made by other scientists as well (Farmer, 1990), but it has received little attention as yet in the social sciences, spatial demographics being an exception. In this field, researchers are beginning to consider migration as a dynamic rather than a static phenomenon, and they are treating spatial demographics as a nonlinear dynamic system of the kind now popularized in chaos and complexity theory. Researchers recognize that behavior depends not only on the rules governing individual migration decisions but also on the locational configuration of interacting populations (Haag and Dendrinos, 1983; Sheppard, 1985).

This research has established three conceptual insights that have relevance to geography and science at large: (1) the stability of any spatial system depends on the nature of the spatial interactions in that system; (2) knowledge of the geographic configuration of a system is significant to understanding its dynamic behavior; and (3) spatial systems with dynamic interactions may exhibit properties of path dependence, considerable sensitivity to initial conditions and external perturbations, and unpredictability over relatively short time horizons.

While these insights can be linked directly to recent arguments in complexity theory, they reflect long-standing concerns in human geography, where there has been continuing criticism of the equilibrium orientation of the theories developed in the 1960s to account for the location of economic activities and settlement systems. Allen Pred's detailed historical research on the evolution of the U.S. urban system (Pred, 1977, 1981), for example, anticipated these conceptual insights, demonstrating how initial advantage, cumulative causation, and interdependencies between cities shaped the system. This demonstrated in practice the ideas of increasing returns and agglomeration that Paul Krugman (1991) has attempted to draw to the attention of economists. Pred's work, together with that of a number of other geographers (cf. Harvey, 1982; Massey, 1984; Scott, 1988a, b; Storper and Walker, 1989; Markusen et al., 1991), has shown how spatial economic processes introduce instability and dynamic complexity, but also path dependence and inertia, into the evolution of any existing economic system.

Geographers have also demonstrated theoretically that spatial dynamics limit the generality of standard economic theories, whether of a neoclassical or political-economic persuasion. They have shown that spatial economies may be highly unstable, that standard theses about specialization and trade and perfect competition may become problematic, and that the free flow of capital between regions need not result in an equalization of profit rates or of access to capital (Webber, 1987; Sheppard and Barnes, 1990). Others have used the conceptual insights associated with nonlinear dynamics to describe more broadly the evolutionary dynamics of settlement systems (Allen and Sanglier, 1979; Dendrinos, 1992).

Similar debates are emerging in research at the microlevel of individual spatial decision making, where standard theories again are dominated by models of spatial equilibrium. For example, recent research into theories of spatial price equilibrium suggests that, in realistic spatial systems, any price equilibrium is at best locally quasi-stable, because some firms are locationally disadvantaged relative to others and because consumers change their pricing decisions in response to price differences (Sheppard et al., 1992). Furthermore, even quite small disturbances from this equilibrium may result in a complex and persistent disequilibrium dynamic of price fluctuation and price "wars."

Example: Form and Function

Another theme of geographic research has been that interactions in space tend to result in—and in turn are affected by—certain regularities in spatial pattern, and geographers have contributed substantially to the multidisciplinary literatures on this phenomenon, particularly as it relates to location theory. One impetus for this research was the observation that a given spatial pattern can result from very different processes—suggesting that function cannot be inferred directly from such patterns.

Just as other disciplines such as physics, astronomy, and biology see patterns as both a reflection of nonrandom processes and an influence on them, geography observes and tries to understand patterns in human settlement and natural landscapes. In part, no doubt, the interest in patterns relates to geography's characteristic use of maps and other graphic displays of information in seeking understanding.

Just as in the case of models of spatial interaction, however, geographers have learned that observable geometries in the social and physical worlds are dynamic in their nature and multidimensional in their explanation. Thus, geographers recognize that in order to understand such dynamic processes it is important to observe them in both time and space. This has stimulated efforts to develop tools for dynamic multidimensional visualization as one way to explore these complex geometries (Dorling and Openshaw, 1992). Geography's curiosity about patterns has stimulated leading scholars to examine patterns in time as well as space and, in turn, how the two kinds of patterns are related (see Sidebar 5.8).

Example: Conflict and Cooperation

Conflicts are rarely confined to one place. They are influenced by developments in other regions, and their effects are usually widely felt. In the ongoing

SIDEBAR 5.8 Long-Wave Rhythms in Transnational Urban Migration

Comprehensive studies of consistencies of pattern and rhythm in economic and political history have shown that a variety of data and their change over time are consistent with "Kondratiev waves" of growth rates, prices, and associated political stresses. Essentially, the explanation is that new technoeconomic systems exhibit a life cycle from innovation to peak activity to replacement and that the expansion and decline of such systems, in succession, stimulate rises and falls in price inflation and other economic forces.

Geographers have shown that such long-wave rhythms can affect spatial flows as well. For example, geographer Brian Berry has shown that global urban growth from 1830 to 1980 displays long-wave rhythmic behavior (Berry, 1991; Berry et al., 1994). By compiling urban growth and migration data for this period, Berry was able to show that the rhythmic behavior was related in part to surges and sags in transnational urban migration; during the same period, domestic rural-to-urban migration exhibited noncyclical trends. This analysis indicated that long-wave historical patterns of economic development have affected spatial patterns of urban growth and that such development has "successively ratcheted global urban growth to new levels of interdependency" (Berry, 1991).

effort to understand the forces of conflict, there is a critical need to consider the relationships among and between places: which places are implicated in particular conflicts and how those conflicts affect different regions and territories. Geography's long-standing concern with identifying, mapping, and analyzing spatial structures and flows speaks to this need. It is manifest, for example, in geopolitical studies that seek to understand how views of territory emanating from different places shape conflict, in studies that explore changing patterns of contact and communication, and in studies that focus on the movement of peoples.

A few examples show the importance of considering such matters in research on conflict and cooperation. Working within the geopolitical tradition, Saul Cohen (1991) has shown how changes in strategic understandings following the demise of the Cold War order have transformed strategic areas of competition—shatterbelts—into gateway regions that link formerly separated territories. Studies in the geography of communication and information have shown how new patterns of connectivity can influence conflict and cooperation (Brunn and Leinbach, 1991). Geographic work on refugees provides direct evidence of the interconnectedness of place, highlighting how flows of people destabilize political regimes and challenge fundamental notions of citizenship and community (Wood, 1994).

Example: Human Health

One of the best illustrations of spatial interdependence can be found in geographic research addressing the spread of infectious diseases. The spread of such diseases is a highly spatial process that can often be understood and predicted by using spatial modeling techniques (see Sidebar 5.9). Research by geographers on the spread of infectious diseases incorporates many of geography's perspectives related to location, synthesis, and scale.

INTERDEPENDENCIES AMONG SCALES

It is impossible to talk about place without reference to scale, and it is impossible to talk about interdependencies between places without considering a variety of different scales. From the earliest times of theory development, geography has been deeply concerned with interdependencies among scales, from global to local. This body of experience is highly relevant for basic and applied science. Relationships between microscale and macroscale phenomena and processes are receiving research attention in many fields of science and are central to knowledge-based questions about such societal concerns as global change.

Attention to interdependencies among scales enables geographers to avoid at least two types of errors. First, the nature of a given phenomenon or process is obscured when it is viewed at the wrong spatial scale. For example, inaccurate or incomplete understandings of local processes and dynamics can result from inferring relationships at one scale based on data collected at another—inferring

SIDEBAR 5.9 Commuting Flows and AIDS

Recent research shows that acquired immune deficiency syndrome (AIDS) infection rates in the New York metropolitan region can be predicted by using measures of "commuting intensity" (Gould and Wallace, 1994). Commuter flows and spatially variable socioeconomic conditions in this region act to channel transmission of human immunodeficiency virus (HIV). The spread of AIDS in the region can be modeled by considering 24 spatial units—boroughs and counties in the region—as the rows and columns of a matrix, whose elements are the average number of daily commuters from each area to all others. From this 24 × 24 "commuter matrix" describing daily flows, a probability matrix (i.e., a stochastic matrix describing a Markov process) is constructed by dividing the flow elements by their row sums. The elements of the eigenvector of this matrix provide a measure of commuting intensity that is hypothesized to be related to the probability of an infected individual infecting another individual.

For each of the three years 1984, 1987, and 1990, the commuter accessibility index predicts AIDS rates with a correlation coefficient (*r*) of 0.85 to 0.92 (see Figure 5.8). In other words, the human structure of the space—rather than the simple geographic space of the traditional map—appears to guide the spread of HIV and its subsequent appearance as AIDS.

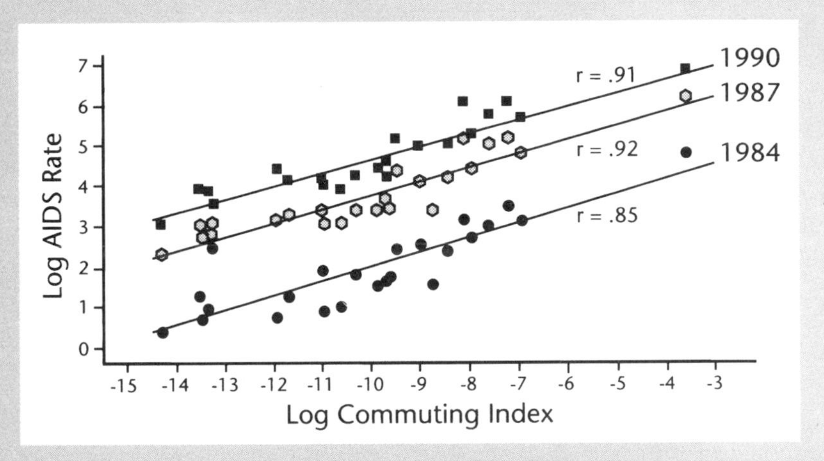

FIGURE 5.8 Scatterplot of log AIDS infection rate against log commuting index for the New York metropolitan area for the years 1984, 1987, and 1992. The best-fitting regression lines and the correlation coefficients for each year also are shown. Source: Gould and Wallace (1994).

subnational trends based on national data (Sidebar 5.8). Second, inadequate attention to scale can result in serious misinterpretations of cause and effect. For instance, an exclusive focus on local scales can lead to explanations in terms of local causes, even when controlling processes occur at regional or global scales (cf. Sidebar 5.10). Likewise, a focus on regional scales of analysis can conceal problems that exist at the local level. Infant mortality rates are exceedingly high in many local areas in U.S. cities, for example, but appear to have fairly uniformly low rates when viewed at the regional level, as is commonly done. Tracing such connections from scale to scale—in particular, examining the importance of processes that operate at intermediate or "mesoscales"—is a significant contribution of geographic scholarship to science.

Relevance to Issues for Science and Society

Across the spectrum of sciences, and increasingly a major focus of geography, is the linkage between macroscale and microscale processes—that is, how phenomena at different time and space scales interact in surprising, disjunctive, and unpredictable ways. Biologists struggle to understand linkages between molecules, cells, and organisms; ecologists between patches, ecosystems, and biomes; and economists between firms, industries, and economies. In these efforts, variants of at least three questions persist: Is behavior of the macrounit of study reducible to the aggregate of microunits? What is universal across scale and what is particular to the scale of analysis? How do agency and structure interact at different scales? Nowhere are these questions more pressing than in the great interdisciplinary questions of origin, organization, and change—in particles, in life, in societies, or in the cosmos.

By not assuming that macroscales are simply aggregates of microscale events and by focusing on mesoscale phenomena to tease out the linkages, geographers help inform our understanding of scale-dependent processes in such diverse science fields as landscape ecology, regional economics, or epidemiology (see, e.g., Sidebars 5.5 and 5.10). In major integrated studies such as those on global change, geographers actively pursue links between global change and local places, enhancing understanding of both scales (e.g., Wilbanks, 1994). Scale relationships are important for scientific understanding of important societal issues such as population and resources, environmental change, economic health, and conflict and cooperation, as shown in the following sections.

Example: Population and Resources

Perhaps no topic evokes more emotion in global change studies than the ultimate human causes of environmental change—the subject of an extended scholarly and public debate. Population and resource use figures prominently in this debate; the "well-known" IPAT identity, representing how environmental

SIDEBAR 5.10 The Importance of Regional Disaggregation

Do national trends accurately portray regional economic activity? Not according to recent work by geographers who have modeled the regional economy of the midwestern United States by the Regional Economics Applications Laboratory of the University of Illinois and the Federal Reserve Bank of Chicago. They show that the Chicago economy restructured at a faster pace than the nation as a whole. Although employment nationally in manufacturing remained relatively flat (around 19 million) from 1970 to 1990, manufacturing employment in the Chicago region declined by almost 50 percent. Output in manufacturing in the Chicago region in constant prices in 1990 was approximately the same as it was in 1970, however, suggesting a 44 percent increase in manufacturing-labor productivity in this period. No doubt some of this productivity increase was due to restructuring of the manufacturing base in the city, as labor-intensive industry was replaced by more diversified and less labor-intensive activities. Chicago led the nation in productivity gains in all but two years in the early 1980s, and it matched or exceeded the productivity gains for Japan in 1982–1988 (see Figure 5.9).

The import-export structure of the region also changed in the 1970s and 1980s. The metropolitan economy became much less dependent on itself as a source of supplies or as a market. To take the analysis to another geographic scale, the Chicago region became even more integrated with the rest of the Midwest. For example, its import of steel from Indiana in 1993 ($2.4 billion) was twice the size of all Illinois exports to Mexico.

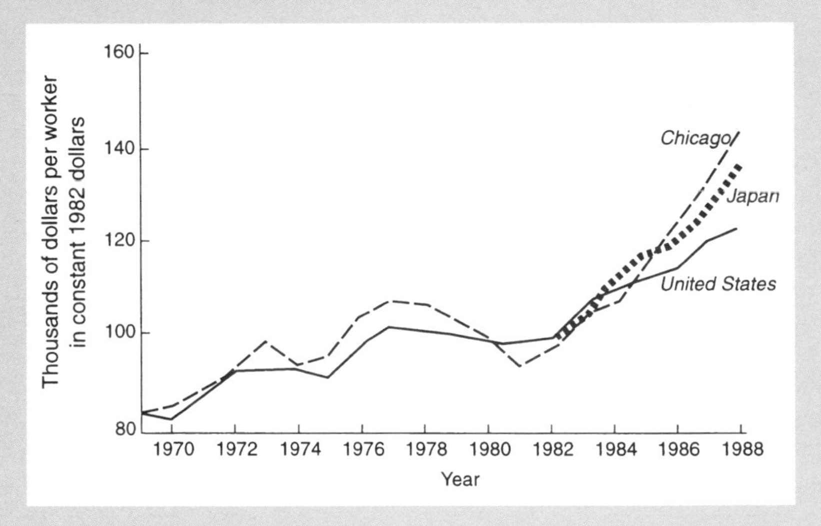

FIGURE 5.9 Manufacturing productivity changes for Chicago and the United States (1969–1988) and Japan (1982–1988). Source: Unpublished graph from the Regional Economics Applications Laboratory, University of Illinois, Chicago (1995).

impact (I) is a consequence of self-reinforcing interdependencies among population (P), affluence (A), and technology (T), is sometimes identified as the controlling process in environmental change, in part because the PAT variables tend to show the strongest associations with atmospheric carbon dioxide and forest and agricultural land cover changes. However, local case studies by geographers often point to a large range of more ''socially nuanced'' factors as the principal triggers of human actions that give rise to trace gas emission, deforestation, and increased cultivation (Meyer and Turner, 1992; Kasperson et al., 1995).

Where global change is concerned, for example, it is clear that some forcing functions operate at a global scale: greenhouse gas composition in the atmosphere and related changes in global climate systems, global financial systems and patterns of control, and movements of technology and information. It is equally clear that most of the individual decisions that underlie economic activities, resource use, and population dynamics are made at local scales. In other words, global processes have impacts on local places, but local actions are the foundations for global trends (Kates, 1995).

Critical questions for science in understanding global change include (1) clarifying the scale(s) at which change should be observed and analyzed and (2) tracing linkages between processes that operate at macro- and microscales. Tracing these connections from scale to scale is a significant contribution to science by geographic scholarship (Blaikie and Brookfield, 1987; Roberts and Emel, 1992; Meyer and Turner, 1994).

Beyond global change per se, geography seeks to identify dynamics between scales for various kinds of resource use and development questions (e.g., Zimmerer, 1991; Bassett and Crummey, 1993; Emel and Roberts, 1995). A particular interest has been in the effects of multinational economic and political structures on regions and localities in developing countries (Watts, 1983; Carney, 1993), especially in areas where ecologies are delicately balanced (see Sidebar 5.11), but similar conditions have been observed in the United States as well (Pulido, 1996).

Example: Environmental Change

The scale of operations plays an important role in deciphering connections among climatic systems. Many of the recent advances in research into global climate change have emphasized the global scale, and the connections among components of the global climate system are now much better understood than they were just a few years ago. From the standpoint of human experience, however, climate is much like politics: it is local. Making the process connection between the now better-understood global circulation patterns and the critical effects they have on small areas (drainage basins of a few hundred square kilometers, for example) has been elusive. Part of the problem is related to computing power and technology, which is stretched to the limit in the simulation of global processes—it is simply not feasible to model local climates in global

SIDEBAR 5.11 Food and Famine in the Sahel

Climate change and markets operate globally through hierarchical systems over which farmer, herder, or district manager have minimal influence. The environmental and social problems these land managers encounter are often well beyond their immediate control, although they may take the blame for the outcomes that follow (Blaikie and Brookfield, 1987). The drought-prone Sahelian region of West Africa is a case in point. People in this region suffer from periodic food crises and, on occasion, widespread and devastating starvation. In the early 1970s the entire region was in the grips of severe famine, and throughout the 1980s, despite foreign aid support, food insecurity was endemic. The Sahel came to be seen as a so-called basket case, a region of structurally induced hunger, declining food output per capita, and a high degree of famine proneness. Insofar as the semiarid tropics are characterized by drought and unreliable rainfall, the Sahelian famine proved something of a test case for understanding the complex relationships between environmental perturbations and catastrophic collapse of food entitlements resulting in mass starvation.

Geographers have reconstructed the history of food crises in the Sahel region, focusing on the dynamics among processes at different geographic scales and the way that these dynamics affect particular places, groups, and classes. This research employed a variety of oral and archival historical sources in combination with ethnographic analysis of social and environmental processes at the community level. For the Sokoto Caliphate (1806–1902) and the colonial and postcolonial periods of north-central Nigeria, for example, the work demonstrated how the integration of peasants into regional and global markets often rendered them increasingly vulnerable to drought-induced harvest failure (Watts, 1983). Famine was not simply the product of colonialism. Rather, market changes exposed some sections of society to the combined volatilities of weather and world markets.

Farmers were in some sense cognitively and practically prepared for variability in rainfall, implementing a standard farm plan every year by orchestrating soil quality, seed varieties, and water conservation practices in relation to the actual distribution of rainfall events. This indigenous practice revealed the capacity of local people to experiment with local resources and to respond to weather variability. However, almost one-third of all rural households were not self-sufficient in food even in normal years. This group of households was especially vulnerable to weather variations and seasonal fluctuations in grain prices. In periods of severe drought, many poor households were forced to liquidate their assets systematically, sometimes resulting in the sale of land and permanent out-migration in search of money, work, and food. Famines thereby intensified existing patterns of social inequality and risk, further polarizing already differentiated communities.

simulations. What is needed is a set of theories that provides rules for connecting a changing global geography of mass and energy with local outcomes.

In ecosystems there is a nested hierarchy of scales so that relatively simple localized assemblages of life forms and their related physical and chemical systems aggregate into larger, ever more complicated associations. Different explanations apply to the behavior and arrangement of the systems at different scales. A riparian forest, for example, adjusts to changes in flooding, groundwater levels, and nutrient loadings in the water and soil. These adjustments are measurable and meaningful within just a few meters in the vertical dimension. At the opposite end of the scale, in biomes—or subcontinental assemblages of ecosystems—these local driving mechanisms are meaningless, and the most useful explanations lie almost completely in the climatological realm. Within a given biome, distributions may be explained best by geologic and landform variables. Successful scientific explanations therefore must start with selecting the controlling variable that is most closely associated in scale terms with the object of study.

Management of environmental change also has important scale considerations. Watershed management in the United States provides an instructive example. Throughout the twentieth century, watershed management has progressively become a federal responsibility. However, the result of national management was a scale mismatch because there are no basins that are truly national in size. Local interests, including resource developers, water and power users, conservationists, and preservationists, have felt isolated from the decision-making process that directly affected them and their watersheds. In the latter part of the century, more localized decision making is becoming common. In Massachusetts, for example, the state coordinates watershed associations organized along drainage basin boundaries. These administrative entities bring together the stakeholders in basins of a few hundred square kilometers to reach compromise solutions in management questions. In the Pacific Northwest, watershed councils of federal, state, local, and tribal representatives operate within basin boundaries to address such problems as balancing economic development and preservation of salmon, objectives that rely on the same watershed resources. The most effective scale for governmental administration of watersheds remains an open question, but the EPA, the U.S. Bureau of Reclamation, the Tennessee Valley Authority, and several other agencies are supporting a National Research Council study of the issue[2] with the ultimate goal of better matching the scales of natural and administrative process.

Example: Economic Health

The economic health of a locality, region, or nation depends on the interaction of processes that operate at many different scales—ranging from global capital flows to local labor markets. Geographers have long been interested in this interplay of global, regional, and local processes—for example, those between global economic forces and local social forces.

Research on economic inequality has revealed that patterns of growth and

---

[2] The study, which is being undertaken by the Water Sciences and Technology Board, is entitled New Perspectives in Watershed Management.

decline are not uniform across nations, regions, or cities. "Third-world inequality" includes rapidly growing small countries, oil-rich countries, and large countries, which seem incapable of breaking out of real poverty. Much like the fractal images of mathematics, extremes of poverty replicate themselves at spatial scales ranging from the global to the neighborhood, implying an irreducible spatial complexity to social irregularity. The heterogeneity across spatial scales reflects variations in political, institutional, and social characteristics and adaptations among places. It also reflects complex processes linking very different scales. Thus, international capital flows link inner-city sweat shops that manufacture clothing in both third-world and first-world economies with affluent and far-flung suburbs and "edge cities" of metropolitan regions.

Differences in economic paths between countries and regions are shaped by differences within those places and also by their differing situations within larger-scale economic and political processes. Within metropolitan areas in many industrialized countries, for instance, suburbanization during the past 25 years has included not only residential development but also the complete range of economic, political, and social activities, with two glaring exceptions: the poorest and least educated households (see Sidebar 5.3) and the highest-order service activities often most directly connected with the global economy. This "spatial mismatch" between the work experience of many inner-city residents and the employment opportunities available nearby has been studied in some detail by geographers and sociologists, including its relationships to processes and policies at regional and national scales.

Example: Conflict and Cooperation

The interest of geographers in scale-related issues involving the connectivity of places is timely because the roles of nation states and localities are undergoing profound change. Developments from both "above" and "below" are challenging the autonomy and power of the state. Internationalization of the economy, development of transport and communications linkages across international boundaries, and growth of substate nationalism and regionalism have pushed scale-related issues of regional formation and interregional interactions to the fore. Although states continue to play powerful roles in many arenas, such issues cannot adequately be addressed using the conventional construct of the state as a discrete analytical unit independent of cross-scale dynamics.

Conflict and cooperation is a good example of a scale-dependent issue that has received recent attention from geographers. Through analyses that look beyond the scale of the state, geographers have contributed to our understanding of the influence of the global economy on local political developments (Taylor, 1993); the nature and importance of cross-border cooperation for the management of social, political, and economic issues (Murphy, 1993); the impacts of global economic restructuring on patterns of interaction (Dicken, 1992); and the influence

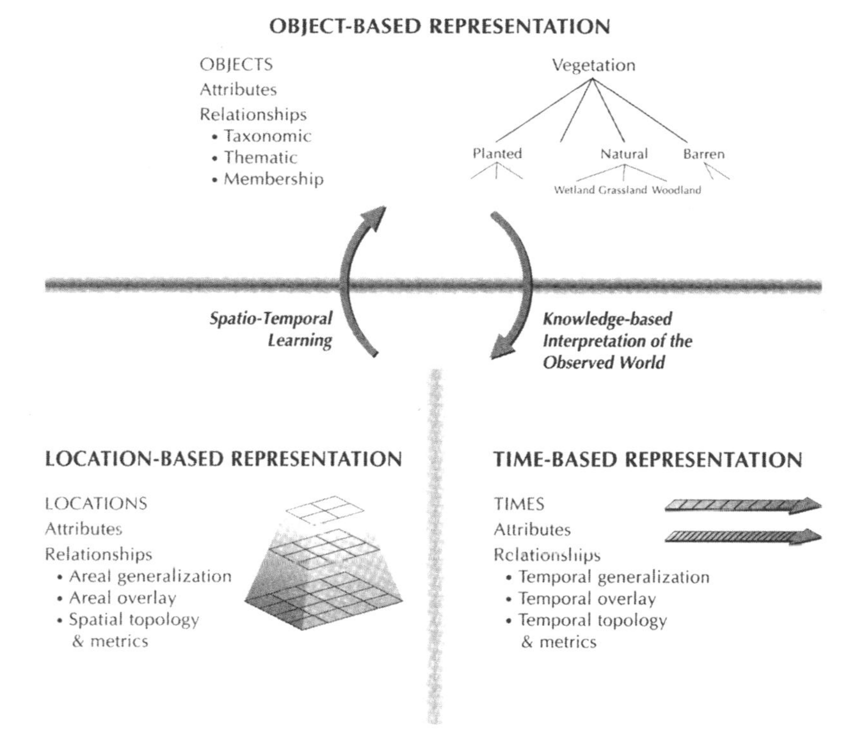

FIGURE 5.10 A spatiotemporal framework in which object-based representations, location-based representations, and time-based representations are treated as complementary approaches to representation of space-time phenomena. This framework is currently applied to design a spatiotemporal data structure that supports research and policy analysis associated with issues of forest succession and management. This integrated view of location, object, and time extends the ability of GISs as a framework for synthesis by considering time (and change) directly, rather than as simply an attribute of space. Source: After Peuquet (1994).

of different social, cultural, and political boundaries on human relatedness (Lewis, 1991).

SPATIAL REPRESENTATION

Many of the substantive contributions by geography to science are rooted in spatial representation. The relevance of geographic research in advancing representational theory and representational tools used throughout science is clear from the widespread interest in GISs and geographic information analysis, but the potential for contribution is far broader. The use of spatial representation as a way to facilitate creative thinking, especially related to nonlinear dynamics, has increased rapidly in the past two decades with the growing decentralized use of computer graphic technologies. Similarly, the use of methods and tools for spatial representation is fundamental to geographic synthesis. GISs, in particular, act as a framework through which information from disparate sources can be integrated and linked to mathematical models and to visual display. Spatial representation has become part of the everyday research experiences of a great many scientists.

Much of the recent geographic research on spatial representation is focused on finding better ways to represent the dynamics of the ''real world,'' for example, by extending fundamental concepts of spatial representation into the temporal domain. Key questions addressed by this research include the following:

1. What is a ''feature'' or an ''entity'' in space-time, and how should these features be classified and coded in digital representation? The development of digital Spatial Data Transfer Standards, a component of the National Information Infrastructure, promises a classification effort with implications for science that rival those of the Linnaean classification in biology.

2. What is the best approach for creating space-time data structures that have a built-in capacity to handle change (in characteristics of spatial or temporal

FIGURE 5.11 Two of Philip Gersmehl's (1990) nine ''metaphors'' for map animation. These metaphors provide conceptual models that map animation designers can use as a basis for building dynamic representations of spatial-temporal processes. The ''stage and play'' metaphor treats a map as a base (or stage) on which action plays out in the form of moving objects and territories particularly suited to representation of human spatial behavior associated with migration or war. The ''metamorphosis'' metaphor was based on the concept of a flexible dynamic object that changes shape over time; it is particularly suited to representation of processes in which area features change size and shape, such as desertification, growth of cities, or the spread of an oil spill.

entities as well as in what constitutes an entity) and the flexibility to support spatial, attribute, and temporal queries; links to process models; and dynamic geographic visualization (GVis) (see Figure 5.10)?

3. Can generalization, spatial filtering, and other geographic "operators" currently applied to spatial data be adapted to space-time data, or is a fundamentally different approach to geographic operators needed?

4. What are the appropriate conceptual models and associated design principles for dynamic display of spatiotemporal data (see Figure 5.11)?

SIDEBAR 5.12 Representing Reliability of Geographic Information

Spatial data support a broad range of research and policy decisions; thus, issues of spatial data reliability (i.e., the quality of spatial data) are central. The National Center for Geographic Information and Analysis, supported by the National Science Foundation, plays a major role in setting the national research agenda through a research initiative (Beard and Buttenfield, 1991) and research challenge on "Visualization of Data Quality" (Buttenfield and Beard, 1994). The winning project in this challenge integrated principles of GVis and exploratory data analysis in the design of an interactive interface for data analysis. The interface allows analysts to monitor spatial and temporal trends in dissolved inorganic nitrogen (DIN) in the Chesapeake Bay as well as to consider spatial variation in the reliability of DIN estimates (see Plate 6).

An alternative to treating reliability as an attribute of data (that can be mapped) is to approach the problem as a question of selecting among a range of possible representations. The more these representations differ, the less reliable any one of them is as the model of reality. An intriguing example of reliability representation that adopts this perspective was developed for application with remotely sensed images. These images result from classification of electromagnetic signals for grid cells (called pixels) that represent square patches of the Earth's surface. The classification procedure involves determining the likelihood that the Earth area represented by the pixel is in each of several possible categories (e.g., vegetation types).

Traditional classified images represent only the most likely category for each pixel (even when the signal processed for that pixel makes classification ambiguous). Multiple images can be used to convey the range of "possible" alternatives to this "best guess." The procedure developed is based on using the fuzzy class memberships (derived through the pixel classification process) as parameters of an error model. The error model generates possible versions of the "truth" (i.e., a version that might result from interpretation by one geographer, soil scientist, or ecologist; see Plate 7). An important assumption in implementing the error model is that the outcomes of neighboring pixels are correlated (i.e., that locations near one another are likely to be similar). In Plate 7, the extent of intrapixel correlation is controlled to produce the four realizations. As the size of this spatial dependence parameter increases, the size of inclusions (i.e., alternative vegetation categories within a region having an otherwise common classification) on the map increases as well (from upper left to lower right in the figure).

SIDEBAR 5.13 Future Geographies

Environmental and societal change will inevitably produce new patterns on the land (i.e., future geographies) associated with unforeseen problems and opportunities. Such patterns, problems, and opportunities may be quite different from today's. A test of geography's new relevance will be its ability to help anticipate, plan for, and improve the future.

Advance warning of the probable magnitude and spatial extent of catastrophic floods is a case in point. Reliable forecasts of the timing, magnitude, and duration of the floods that occurred in the U.S. Midwest in 1993, in Arizona in 1994, and in parts of southern California in 1995 would have enhanced preparedness and saved money and lives. Flood characteristics are quite complex. They result not only from heavy precipitation events but from antecedent soil moisture and land surface conditions as well. Often, human modifications of the land also are important. Changes in weather patterns that might accompany climate variability or change make forecasting large floods particularly problematic.

Accurate forecasts of such future geographies are not currently possible. Preliminary efforts to anticipate catastrophic floods by Hirschboeck (1991) and colleagues at the University of Arizona, however, are promising. They illustrate how evaluations of atmospheric circulation anomalies, in conjunction with assessments of land surface conditions, can improve forecasting skills significantly. They also show that statistical considerations of rainfall and stream discharge observations alone are insufficient to reliably forecast truly large floods. New approaches based on climatic and geographic analyses are suggested and illustrated by showing how the 1993 flood in the U.S. midwest and the winter 1994 flood in Arizona could have been anticipated. In addition to both episodic and persistent behaviors in atmospheric circulation regimes, including mesoscale anomaly patterns, basin-wide soil saturation stemming from persistent circulation patterns is one of several observable factors (e.g., see Figure 5.12).

Geographers have also attempted to forecast global food resources (potential grain yield) under several plausible climate change scenarios (e.g., Rosenzweig et al., 1995). Based on three global climate models, their results suggest that possible future climates (with twice the current levels of carbon dioxide in the atmosphere) will cause decreases in global food production, although the geography of the changes will be uneven (see Plate 8). With major adaptations in agricultural practices at the farm level, estimated decreases in productivity should be reduced, but global productivity still will be below the current level. Although these future scenarios are imprecise, even about the timing of such changes, they are among the first steps necessary to develop meaningful forecasts of our future world.

Nearly two decades ago geographers were addressing the question of whether the potential biomass productivity of the Earth could be estimated (Terjung et al., 1976). Although such an estimate obviously depends partly on the advance of technology, even with optimistic estimates of technological change, various analyses have suggested that the Earth might be able to support only about 4.5 billion people on a high-protein diet (the U.S. Bureau of the Census estimated the 1994 world population to be about 5.6 billion people). This was not intended as a final answer to the question, but it indicates the seriousness of future-oriented research on issues related to sustainable development.

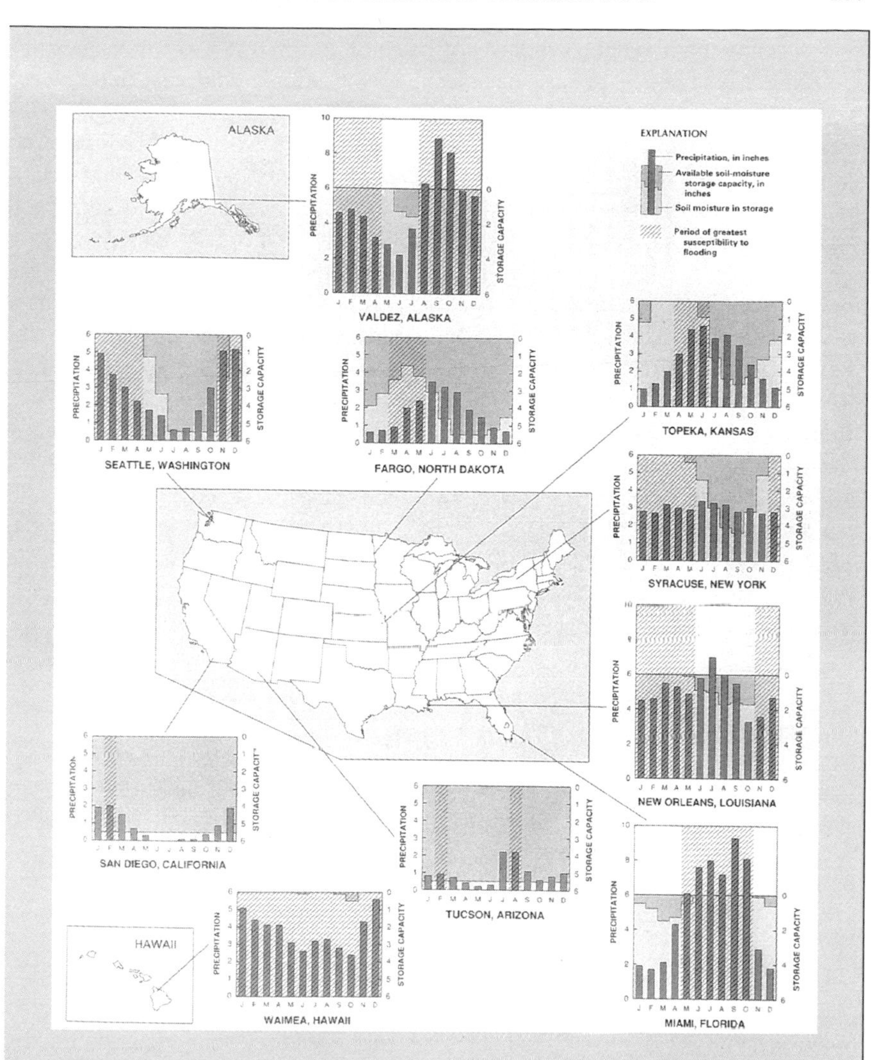

FIGURE 5.12 Effects of average monthly precipitation and soil moisture on susceptibility to flooding at selected locations in the United States. The dark bars on the histograms show average monthly precipitation at selected locations. The lines on the histograms indicate available soil-moisture capacity. The hachured areas on the histograms indicate the times of greatest susceptibility to flooding. Source: Hirschboeck (1991).

A simple but dramatic synthesis of research directed to these four questions is a video developed by geographers at the U.S. Geological Survey. In the video, topographic maps, satellite imagery, land-use maps, and digital terrain models are linked in a dynamic depiction of urbanization in the San Francisco Bay Area between 1850 and 1990 (see Plate 9). The same representational tools are currently being linked to a prototype urban growth model so that analysts can investigate what human occupance of the bay area might look like in 200 years.

Geographic research contributions to the science of spatial representation include work on the classification of geographic entities and visual representations of data reliability (see Sidebar 5.12). Geographers are also linking cognitive and digital representations of space—for example, as part of an interdisciplinary effort to understand the interaction between human spatial cognition and way finding. One component of this research focuses on visually impaired populations (Golledge, 1991), work that has implications for the guidance of robotic vehicles. Additional basic research has addressed such issues as scale effects in spatial cognition, ways in which orientation and direction information is handled in memory, development of configurational (i.e., maplike) understanding, and the ability of visually impaired people to use configurational knowledge to determine shortcuts or take detours around obstructions.

REFLECTIONS ON GEOGRAPHY'S CONTRIBUTIONS TO SCIENCE

By way of example, this chapter has illustrated how geography's perspectives and techniques contribute to understanding key issues in science and strengthen what science offers to the resolution of critical societal problems. The potential for further contributions is significant. For instance, a powerful tool for integrating a variety of dynamic processes to anticipate possible futures is the description of "future geographies"—maps of evolving patterns of change, related to real places and the concerns of those who live there (see Sidebar 5.13).

If geography is to increase its contributions to scientific understanding, however, both geography and the other sciences need to develop more productive partnerships that combine their unique perspectives and approaches to problem solving. Geography itself needs to be engaged more often in research activities that embrace and pursue broader contributions to science, at least partly by showing a greater concern for critical research problem definition by the larger research community. The family of sciences, in turn, needs to be better informed about geography and how its perspectives can contribute to scientific understanding. Both of these priorities call for increased interactions between geographers and their counterparts in other sciences: increases in quantity, quality, diversity, and orientation to critical issues.

6

Geography's Contributions to Decision Making

Geography's long-standing concern with the evolving spatial organization and material character of the Earth's surface is of great relevance to decision makers in business and government. Whether the issue at hand is the location of a new public facility or the development of a stream restoration project, decision makers must consider such geographic matters as location, the relationship between processes at different scales, and the changing character of particular environments and landscapes. As such, geographic expertise can be of great importance in helping organizations and individuals operate more efficiently and make better-informed decisions.

Geographers contribute to policy and decision making in a variety of ways (Wilbanks, 1985). One contribution is through the publication of research findings in professional journals and other open-literature outlets. These insights usually influence decisions indirectly. Scholarly publications tend to influence society's general "climate of understanding" about issues, and society's opinions are transmitted to decision makers through a variety of channels (Weiss, 1977). Although most geographers believe that published research is more valuable before decisions are made, it is often more visibly used in justifying decisions made on other grounds, at least in public policy making (Wilbanks and Lee, 1985). One area of contribution in geography, incidentally, has been in understanding how geographic circumstances fundamentally influence decisions made by public policy makers (e.g., Clark, 1985; Murphy, 1989; Wolch and Dear, 1993).

A second kind of contribution is through reports produced for specific users. Particular problems and questions are posed—usually involving the application of available knowledge rather than advances at the frontiers—and the answers are delivered on a schedule within a predetermined budget, in situations where the timing of a contribution is crucial to its impact. Most of these reports become a part of what has been called a "fugitive literature," seldom peer reviewed (though often intensely scrutinized) or cited in computer-accessible bibliographic databases, but directly impacting public and private decision makers. Although such professional work is usually associated with consulting firms and other nonacademic institutions, it is also a staple of "soft money" research centers in universities. Many leading geographers have been influential in this way, without always being identified explicitly as geographers.

A third contribution, and often the most powerful, is when geographers become a part of the decision-making process, interacting on a personal, confidential basis with decision makers, drawing on a combination of formal knowledge, professional judgments, and mutual trust and effective communication. These roles are seldom reported in the published literature; in fact, the contribution often depends on maintaining confidentiality and letting others take the credit, with satisfaction derived from seeing the right kinds of policy decisions made. Many geographers, from Gilbert White and Edward Ackerman to William Garrison, John Borchert, Harold Mayer, and Brian Berry, have shaped policy in this way, but many of their accomplishments are not reported in the literature.

The problem for this chapter is reporting how geography contributes to decisions when the most powerful impacts are often the least documented—and the least documentable. Given the constraints inherent in the complicated relationship between science and government on the one hand and academia and business on the other, the committee has tried to emphasize subjects rather than specific impacts, offering a mix of evidence based mainly on the open literature but also referring in some cases to personal contributions that extend beyond formal publications. The committee has also tried to highlight issues where it believes geographers should be contributing to well-informed decisions but, for a variety of reasons, are not now contributing in significant ways.

The next section of this chapter discusses the various decision-making "arenas" in which geographers work. The following sections illustrate geography's contributions at several scales: regional and local, national, and international.

ARENAS FOR DECISIONS

Geographers and geographic perspectives have found important application in decision making in both the private and the public sectors. Geographers serve the public sector in many different roles, as government employees, consultants, private citizens, and volunteers for public advisory boards at levels from local to international. Private-sector companies frequently use geographers and geographic knowledge to make location, routing, and marketing decisions and for the management and analysis of spatial information in support of business decisions and communications at a variety of scales, from local to international.

The role of geographers in private-sector decision making is growing rapidly, with improved technologies for decentralized geographic information systems (GISs) use and increased access to georeferenced information, and these roles are becoming strategic as well as operational. A wide range of private companies use geographers and geographic perspectives in their locational decision making. These include retail marketing chains (e.g., Dayton-Hudson, a major retail firm headquartered in Minneapolis), railroads (e.g., Southern Pacific Railroad's land division), electric power and gas utilities, international import-export firms, transportation and travel service organizations, publishing firms, and real estate planners and investors.

REGIONAL AND LOCAL DECISIONS

Cattle ranchers in Brazil hire bulldozers to clear tropical rainforest for cattle ranching; farmers in Kenya build terraces to fight soil erosion on sloping cropland. The first set of decisions leads to environmental degradation, the second to environmental conservation. Steel mills close in the mature industrial regions of the developed world; semiconductor production moves to the newly industrializing countries of Malaysia and Thailand; high-technology firms spring up along the M4 Corridor west of Central London and in Silicon Glen near Glasgow, Scotland; major centers of retail and office activity known as "edge cities" spring up where major highway intersections occur beyond the suburbs of the 1970s, while blocks of apartments and townhouse communities locate close to subway systems, such as the Metro in Washington, D.C. The outcomes of these regional and local decisions affect the well-being of people, alter the look of the land, and set up new geographic patterns that affect the next set of location decisions that people make.

At local and regional scales, geographers assist decision makers by providing information and analyses related to such issues as the management of hazards, management of complex urban systems, and resource allocation, often wrestling with their overlapping roles as scientists and citizens. In addition, geographers advise local and regional government agencies about the design of geographic databases and the use of GISs. The following sections illustrate a range of contributions.

Urban Policy

Cities themselves are functional regions connected with other places by networks of transportation, communication, finance, and trade. Their internal structures can be distinguished according to such characteristics as race and ethnicity, housing, business activities, industrial processes, natural resource consumption, and pollution potential. The urban ghetto, for example, is at once the result of social, political, economic, and geographic processes, and addressing any one or two of these processes alone will not suffice to improve living conditions (Rose, 1971). Effective urban policy making requires an understanding of these spatial and functional characteristics, which are strongly geographic in nature (e.g., see Sidebar 6.1).

Metropolitan areas consisting of constellations of cities pose special problems and opportunities for policy makers owing to their political, industrial, and social complexities. The Los Angeles metropolitan area, for example, has evolved into industrial agglomerations based in part on high-technology firms that specialize in production for military applications (Scott, 1993). With the recent declines in defense spending, regional policy makers have worked to decrease the region's reliance on this sector of industry. Based on geographic analysis of the region's economic structure and role in the world economy, the Los Angeles County Metropolitan Transportation Authority and other agencies are developing an advanced ground transportation industry to take advantage of the region's skilled labor force and manufacturing capabilities.

Cities, of course, are more than built environments. People are their primary components, and significant inequities exist among city dwellers with respect to the basic necessities of life. Urban policies for the homeless, for example, show sharply defined geographic components (Dear and Wolch, 1987). Social and economic polarization, cyclical unemployment, changed housing and investment policies, and government policies for deinstitutionalizing the mentally ill have contributed to the increase in homelessness. Tolerance for the homeless tends to decline as distance from the city center increases. Policy makers and volunteers often reinforce the geography of the problem by concentrating their efforts to reduce homelessness on the central city. The policy implication is that homelessness, although rooted in the fabric of general society, has become a "city problem," and it is in the city that it will have to be resolved.

Water Resources

This nation's water resources are managed through massive public investments. By controlling the supply and distribution of water, reducing flooding, and offering recreational opportunities, the United States has created a partly artificial and partly natural hydrologic system that supports economic development, raising the quality of life for many but also decreasing the quality of life for others.

The diversion of water from rivers for irrigation, industrial uses, and urban water supply reshapes the character and geographic distribution of water-dependent ecosystems, sometimes eliminating them altogether. Overdrafts of groundwater supplies have lowered the water table in many areas, particularly in the western states, further altering surface ecosystems dependent on shallow groundwater.

SIDEBAR 6.1 Geography and Urban Policy: Housing in Minneapolis–St. Paul

In the late 1960s urban geographers established that housing submarkets in American metropolitan areas operate on the basis of geographically defined sectors (Abler et al., 1971). New housing built on the suburban fringe sets in motion waves of housing vacancies that move inward through these sectors toward the center of the city. Households relocate outward as vacancies move inward. In addition, the oversupply of housing in one sector can produce soft markets that lead to newcomer concentrations, while leaving housing markets in other sectors unaffected.

This research has been used by the Metropolitan Council of the Twin Cities to formulate housing policy for the Minneapolis–St. Paul metropolitan area. The council subdivided the metropolitan area into several housing sectors and monitored changes in the number of households, construction, demolition, and housing prices in each sector (see Figure 6.1). The council discovered that the oversupply of housing in some sectors, stimulated in part by overly permissive and excessively generous development controls and utility extensions at the suburban edges, caused difficulties for certain central-city neighborhoods. Adjustments in development controls provided the council with an improved method for dealing with the effects of migration and vacancies in city neighborhoods.

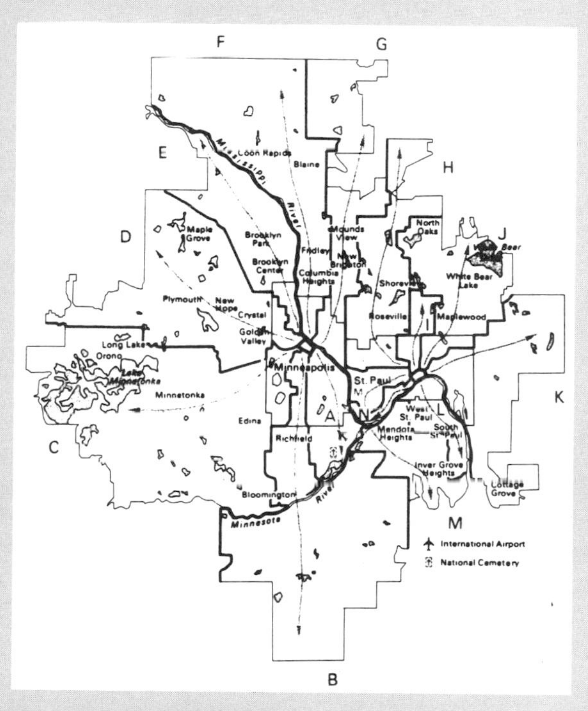

FIGURE 6.1 Housing market sectors in Minneapolis–St. Paul. Sector boundaries are shown by heavy solid lines, and sectors are denoted by capital letters A to N. Source: Adams (1991, p. 113).

Commonly accepted perspectives on the competition for use of public resources, such as Garrett Hardin's "tragedy of the commons" (Hardin, 1968), view water resource allocation issues in terms of economics. An equally valuable perspective is geographic: water resource management problems are problems of geographic distribution. Consumers of water are concentrated in urban areas or in regions of fertile soil, whereas the water sources to sustain them are dispersed. New York City, for example, maintains an elaborate system of reservoirs and water treatment plants that extends more than 600 miles from the city. The Colorado River supports 20 million people, most of whom live far from the river in southern California, eastern Colorado, northern New Mexico, and central Arizona. Geographers contribute to the successful management of such far-flung systems by bringing their geographic perspectives to bear on the analysis of regional impacts of water resource decisions (e.g., see Sidebar 6.2).

The various sectors of American society are unequal consumers of water. In 1990, 48 percent of the total withdrawals from surface and groundwater supplies went to thermoelectric power generation, 34 percent to irrigated agriculture, and 7 percent to industry. Only 10 percent of all withdrawals went to the public water supply—the consumers who are bombarded by conservation messages when shortages occur (Sloggett and Dickason, 1986). Any national strategies for water conservation will have unequal consequences for various sectors and parts of the country, as well as locally and regionally.

In water resource management those who benefit are not necessarily those who pay. The federal government subsidizes western agriculture through irrigation development and eastern industry through flood control. In the generation of hydroelectric power, populations close to the river that generates the power, or populations who use the river, sometimes pay significant opportunity costs in terms of alternative uses they must forgo so the river can generate the electricity that benefits distant consumers. The geographic distribution of who benefits and who pays offers insights to decision makers who must balance competing social demands with a limited resource.

Geographers contribute to successful water management by providing data and analyses to help policy makers reach sound policy decisions. For example, geographer Edward Fernald established the Florida Resources and Environmental Analysis Center to collect and disseminate information to help solve statewide problems. The center has prepared state reference atlases, including the *Water Resources Atlas of Florida.* It also maintains data on flood-prone areas, inventories the uses of state-owned lands, and assists state and local areas in applying GIS techniques. California geographers have also produced a water atlas (Kahrl, 1979).

Retail Marketing

The old adage about the three most important factors in retail success—location, location, and location—still rings true. Location, of course, can refer

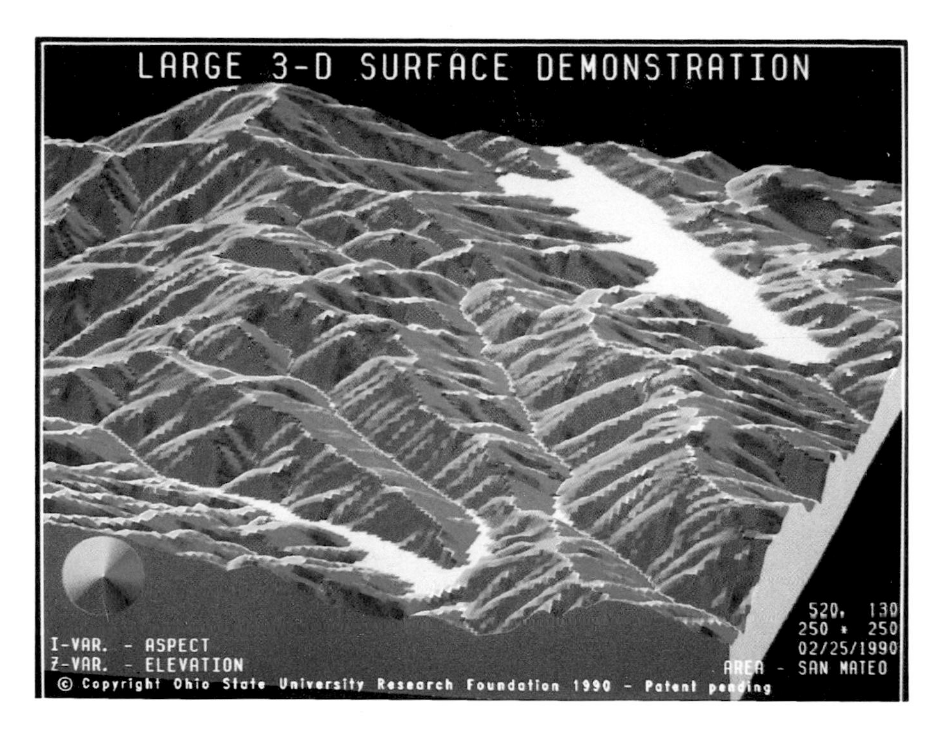

PLATE 1. Three-dimensional map of the San Mateo area of California near the San Andreas fault. The lake in the background lies on the fault. The map uses 128 colors to enhance perception of relief and was generated by using the patented (U.S. Patents 5,067,098 and 5,283,858) MKS-ASPECT™ process described by Moellering and Kimerling (1990). Copyright 1990 by the Ohio State University Research Foundation.

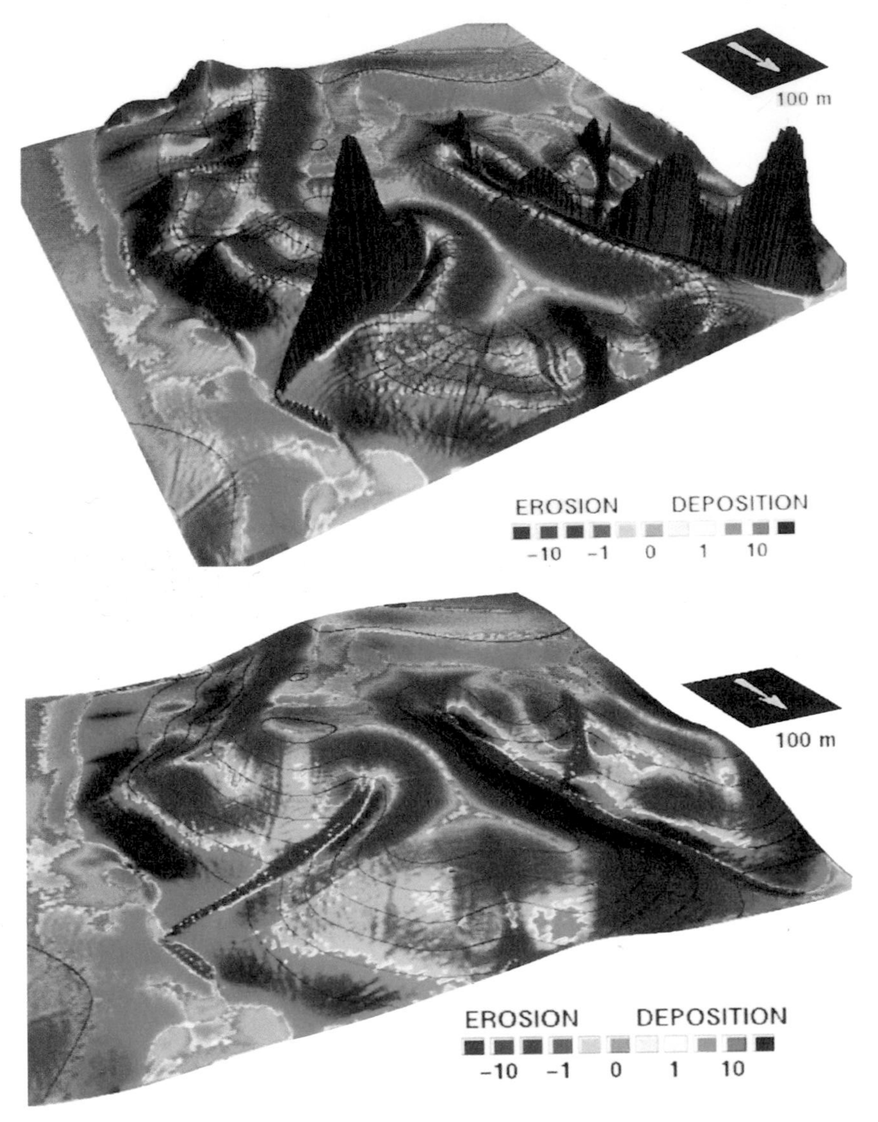

PLATE 2. *Top*: Visual representation of a mathematical model of topographic potential for erosion and deposition of a terrain. Color represents the transport capacity of the surface showing its directional derivative (topographic potential for erosion and deposition). Potential for erosion is high where sediment transport capacity increases (red areas). Potential for deposition is high where transport capacity decreases (blue areas). *Bottom*: Three-dimensional view of an actual terrain that has been smoothed and vertically exaggerated. Colors represent the potential for erosion and deposition. The two images together allow evaluation of the relationship between the shape of the terrain and the results of the erosion/deposition model. Source: Mitasova et al. (1996).

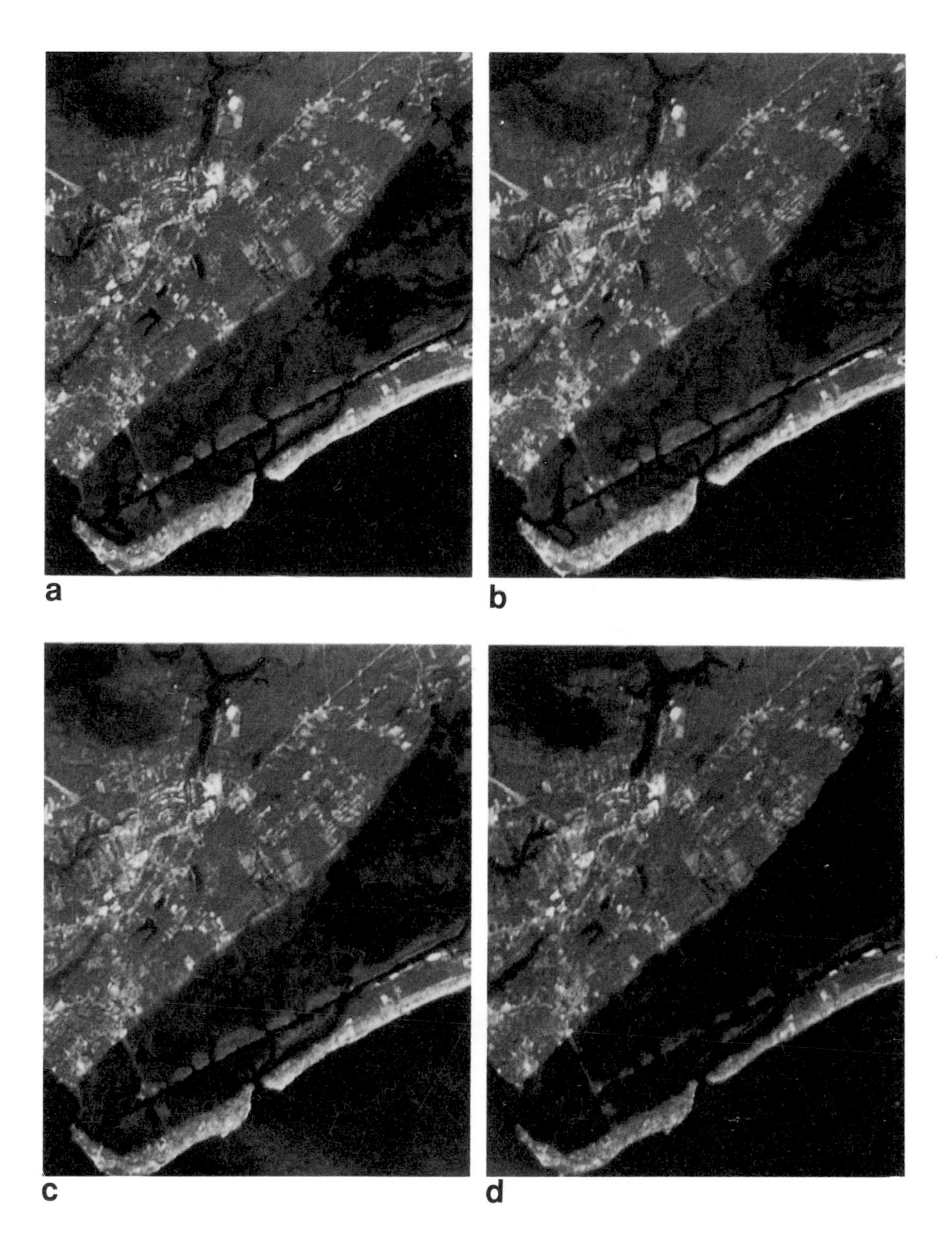

PLATE 3. Projected effects of a sea-level rise on a coastal region near Charleston, South Carolina, based on a combination of Landsat Thematic Mapper (TM) data, digital files of elevation and bathymetry, and data on land cover. The figure shows year 2100 sea-level rise scenarios for (a) a baseline estimate of a 28-cm rise above present mean sea level (MSL); (b) a low-end estimate of a 3-cm rise above present MSL; (c) a best estimate of a 46-cm rise above present MSL; and (d) a high-end estimate of a 124-cm rise above present MSL. Inundation is shown by dark shading. The low-lying populated areas would be inundated under high-end estimate conditions. Source: Jensen et al. (1993b).

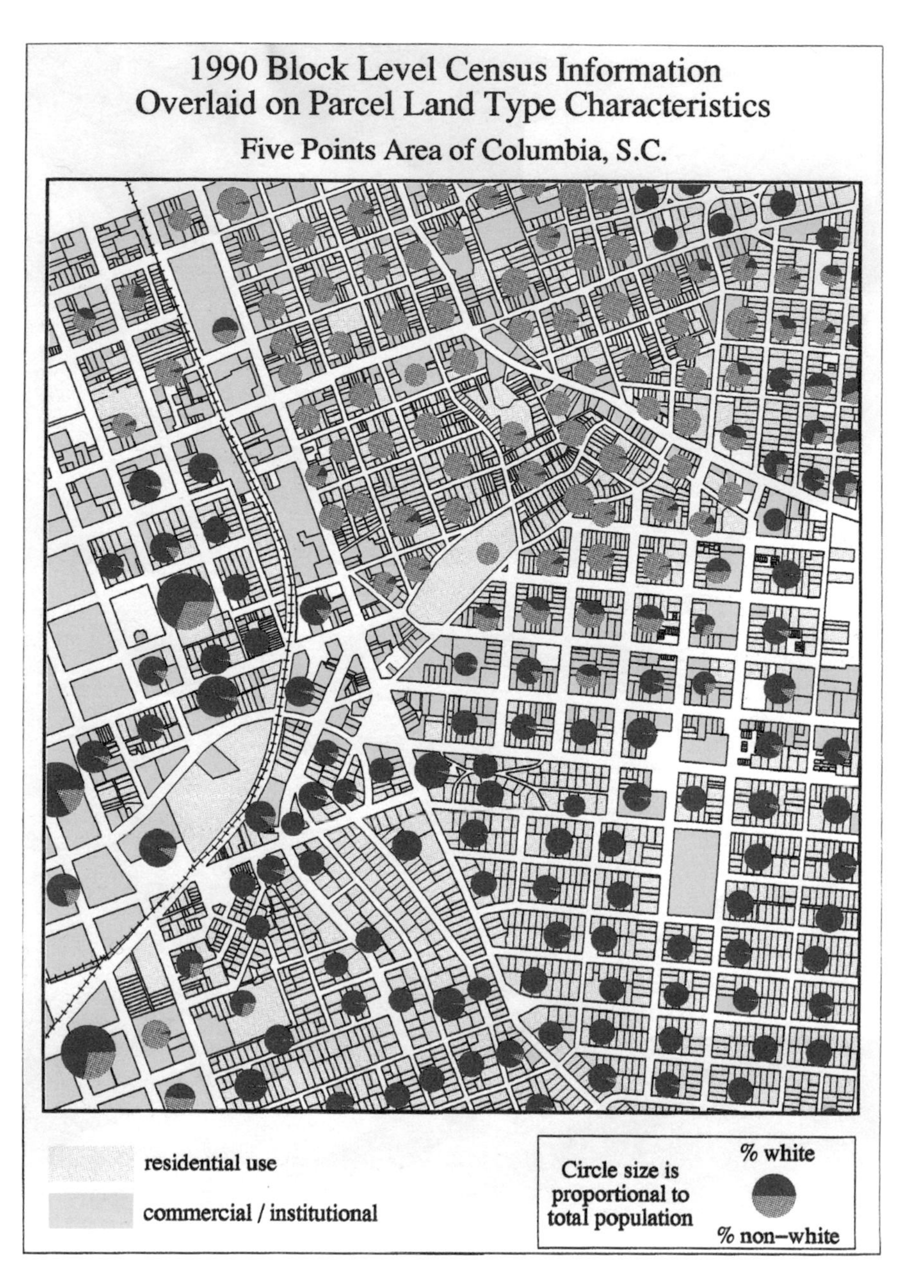

PLATE 4. Population size, racial composition, and land use in a Columbia, South Carolina, neighborhood. Source: Cowen et al. (1995).

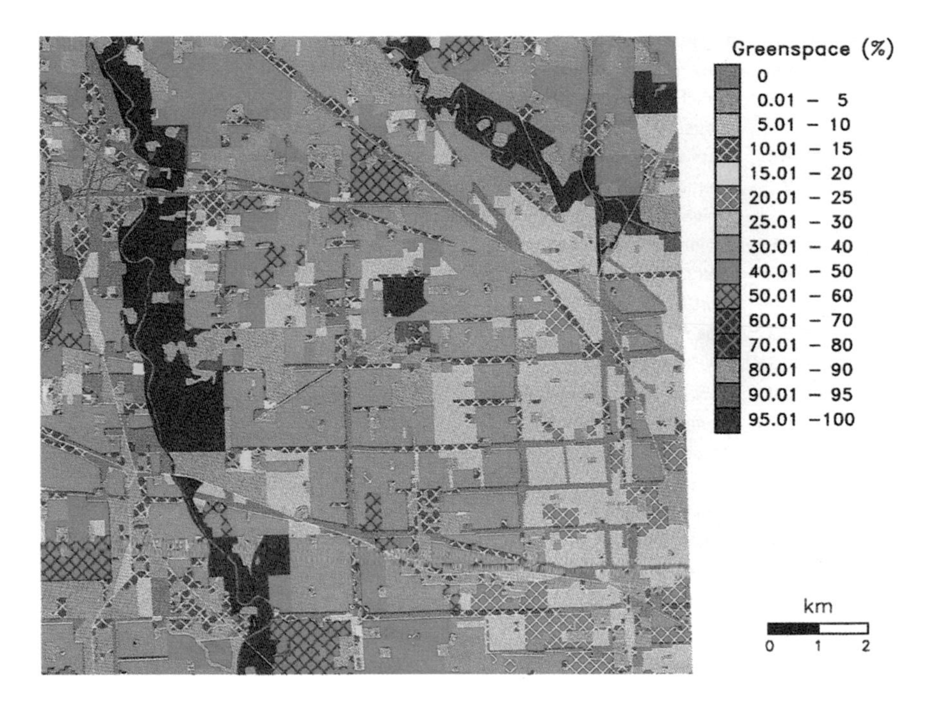

PLATE 5. A map showing the characteristics of land cover through the measurement of greenspace (i.e., trees, shrubs, and grass) in Chicago, Illinois. Source: Grimmond and Souch (1994).

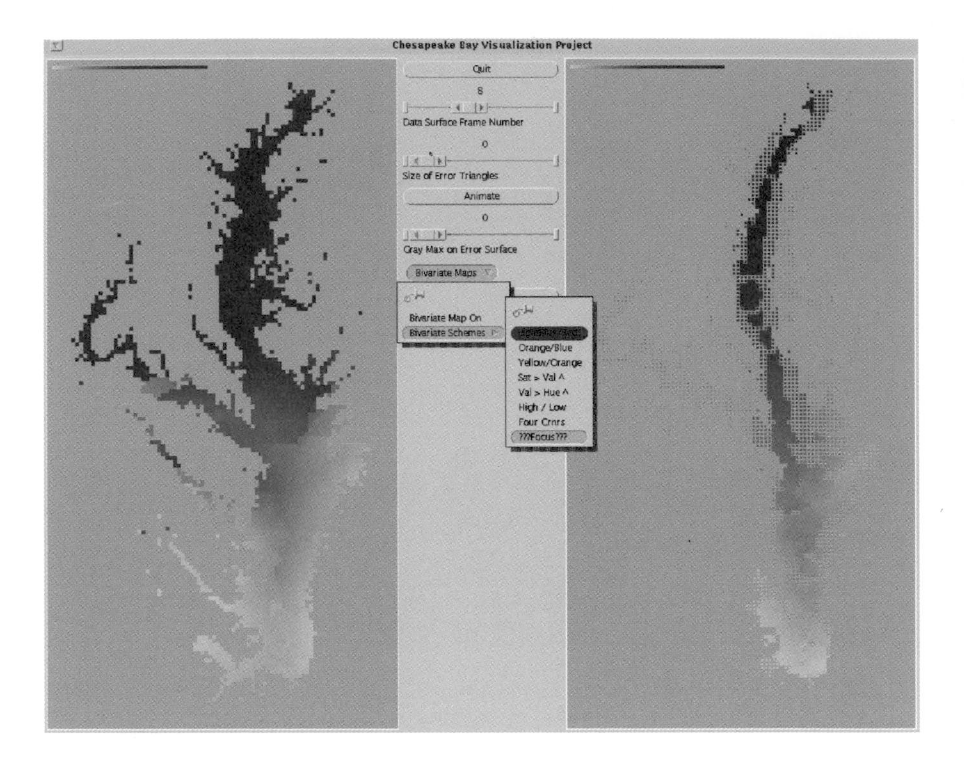

PLATE 6. The left panel of the interface shows the dissolved inorganic nitrogen (DIN) surface interpolated from values at 49 sample points in the Chesapeake Bay (dark red represents high DIN concentrations). The right panel depicts the same data filtered according to the reliability of estimates generated by spatial interpolation. Grid cells for which the value is highly reliable are completely filled. As cell estimates become less certain, the proportion of the cell filled with a shade of red decreases to the point at which the DIN level of highly unreliable cells cannot be interpreted at all. Source: MacEachren et al. (1993).

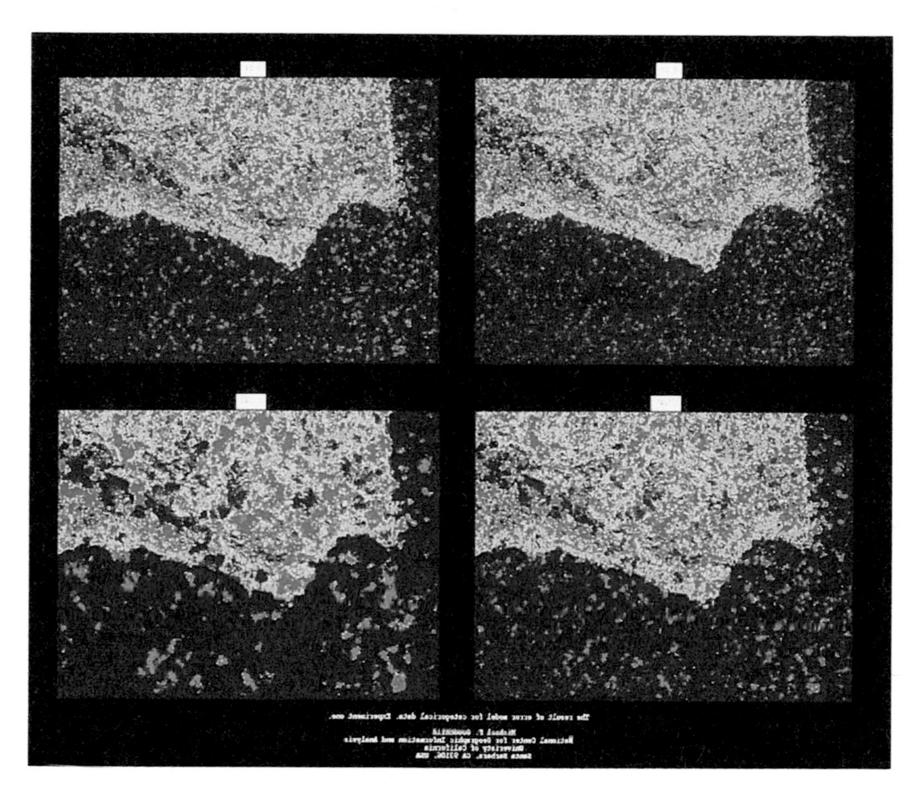

PLATE 7. Four "realizations" of vegetation distribution in the Santa Barbara, California, area. Each represents a classification of grid cell by grid cell (i.e., pixel by pixel) reflectance signals received by a satellite sensor (the Thematic Mapper). Each view is a possible interpretation of what the Thematic Mapper has measured for each cell. The difference among the possibilities results from manipulation of a weighting parameter that adjusts the extent to which each pixel's immediate neighbors are taken into account in assigning a pixel to a vegetation class. High spatial weighting produces more homogeneous regions with few isolated pixels (by assigning pixels whose reflectance puts them near a class boundary into the class most similar to that of their neighbors). Source: Goodchild et al. (1994, Plate 20).

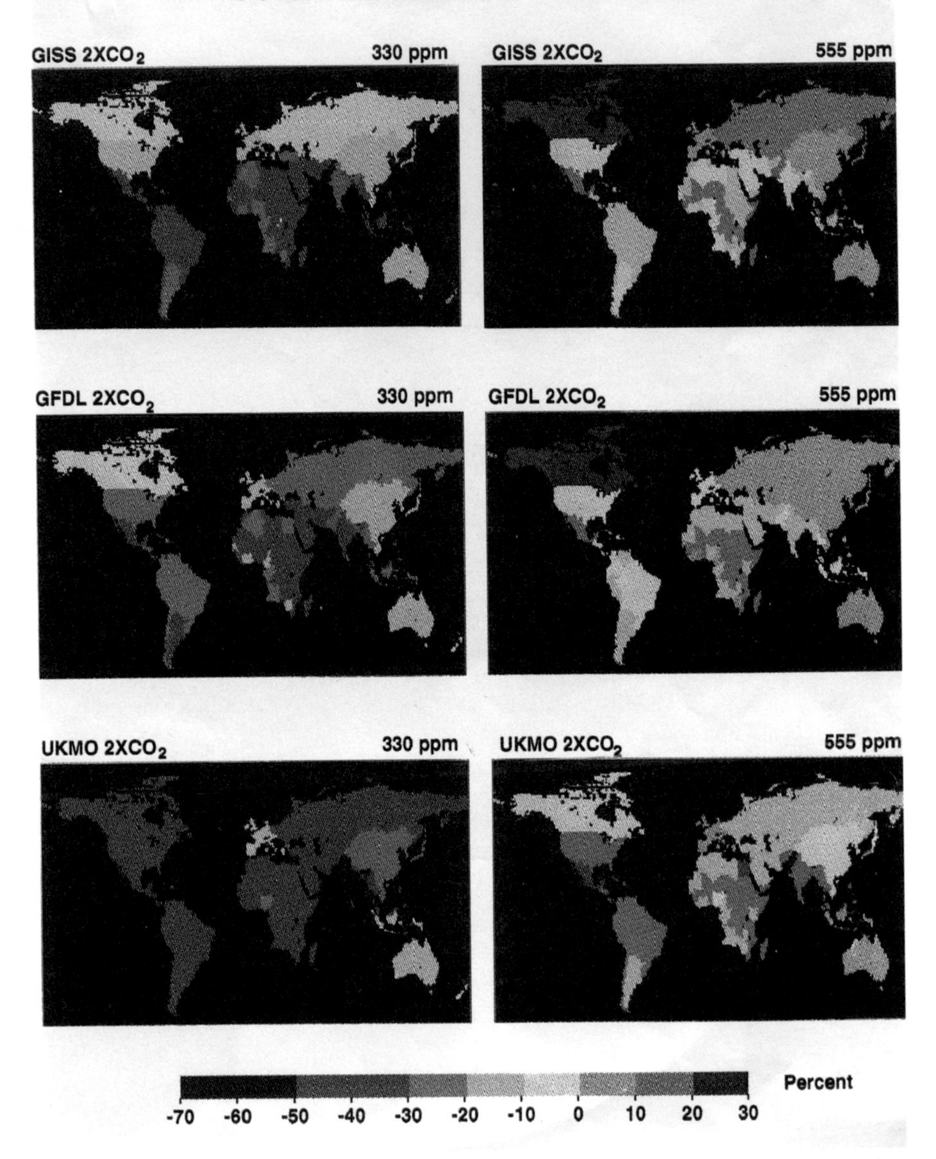

PLATE 8. Computer simulations showing the effects of climate change resulting from a doubling of atmospheric carbon dioxide on national grain (wheat, rice, coarse grains, and protein feeds) yields. The colors show the changes (with respect to the present) in grain yields under climate changes predicted by three general circulation models: GISS, Goddard Institute for Space Studies model; GFDL, Geophysical Fluid Dynamics Laboratory model; UKMO, United Kingdom Meteorological Office model. The left column isolates the effects of climate change on grain yields. The right column shows the combined effects of climate change and doubled of carbon dioxide levels on grain yields. Source: Rosenzweig et al. (1995).

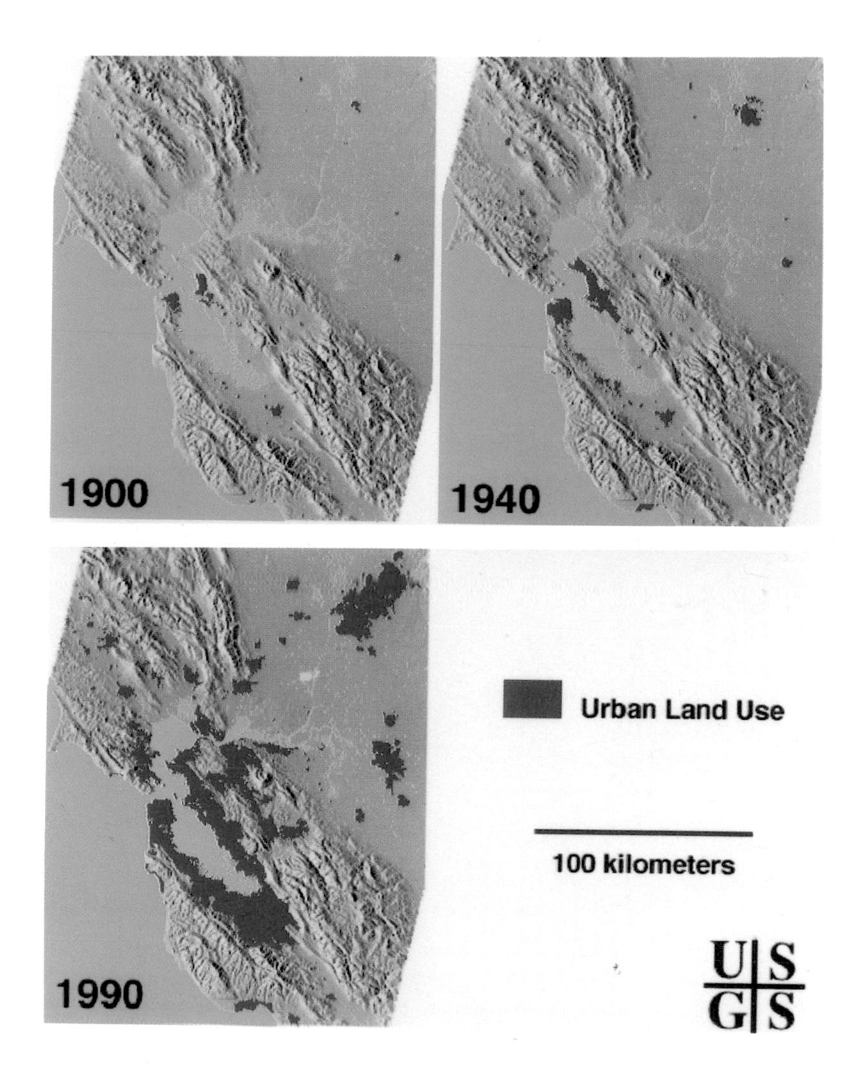

PLATE 9. Images depicting urbanization in the San Francisco Bay area for three time periods: 1900, 1940, and 1990. The images were prepared from map (1900, 1940) and Landsat (1990) data. Source: William Acevedo and Leonard Gaydos, U.S. Geological Survey, Ames Research Center.

PLATE 10. Satellite images of the lower Mississippi River Valley near Glasgow, Missouri. North is to the top of both images: (A) Image taken in September 1992 (prior to 1993 flooding). The active river channel (black) meanders within the floodplain, which is under active cultivation. An arcuate feature in the upper left is an oxbow lake. (B) Image taken of the same area during the 1993 flood. Flooding inundated the entire floodplain in the lower two-thirds of the image. Terraces on the northeast side of the river escaped flooding in the upper part of the image. Source: SAST (1993).

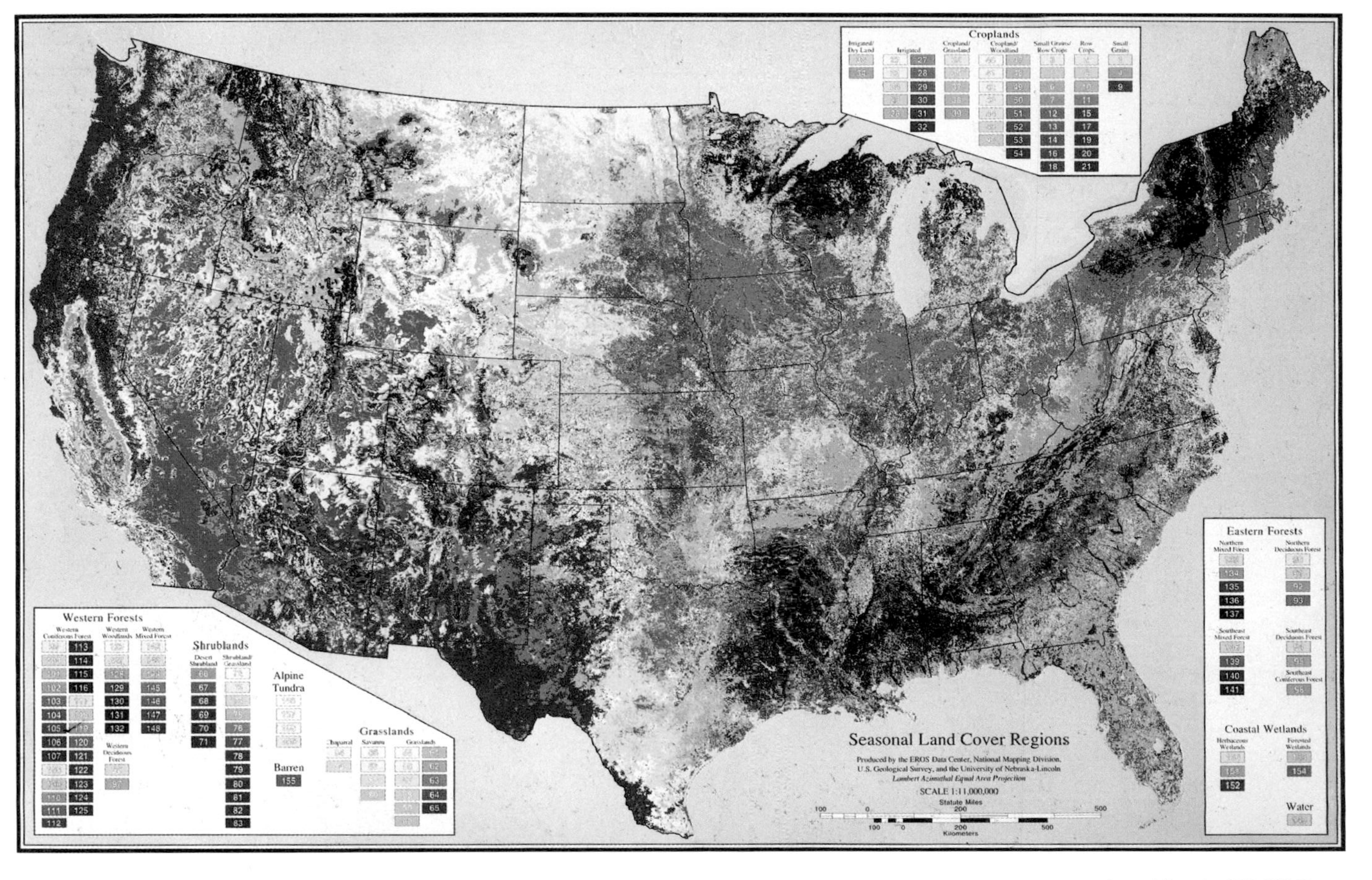

PLATE 11. A 159-class map of seasonal land cover regions of the conterminous United States. The map was produced by the EROS Data Center, National Mapping Division, U.S. Geological Survey, and the University of Nebraska. Source: Loveland et al. (1995).

SIDEBAR 6.2 Geography and Management of the Colorado River in the Grand Canyon

The flow of the Colorado River in Grand Canyon National Park is controlled by releases of water from Glen Canyon Dam, a structure governed by a set of policies known as the "Law of the River." This collection of laws and operating rules specifies that the dam will be operated to deliver particular amounts of water downstream, generate electrical power, store water upstream, and trap sediment. In Grand Canyon National Park, which is located downstream from the dam, these operating rules have resulted in unacceptable erosion of riverside beaches, destruction of habitat for endangered fish, and creation of hazardous conditions for recreational users of the river.

Research by Schmidt (1990), Bauer and Schmidt (1993), and other geographers showed how sediments in the river were being exchanged between beaches and deep pools in the river. Under flow conditions that had occurred naturally before the dam was closed, floods occasionally scoured the pools and built the beaches. Researchers determined that the dam could be operated to occasionally stimulate such flows. They further estimated that some sand still entered the Grand Canyon from undammed tributaries and that there was still enough water flowing through the canyon to distribute this sediment. After extensive congressional hearings, including testimony by Schmidt and other investigators, Congress passed the Grand Canyon Protection Act of 1992. This act directed the U.S. Bureau of Reclamation to operate the dam in a manner that protected downstream resources, including beaches and habitats they supported.

The Bureau of Reclamation operated the dam experimentally for a year to determine the outcomes of various policies. In March 1996 the bureau released from the dam a modest flood flow of 45,000 cubic feet per second for a week. The flow scoured sand from the floors of pools in the canyon and deposited it on reformed beaches and bars, mimicking natural floods that were common before closure of the dam. The beaches and bars restored lost habitat for vegetation and fish and stabilized campsites for the 20,000 river runners who use the canyon each year. The success of the experimental artificial flood suggests that it will become a once-in-every-five-years feature of the dam's operation in a continuing effort to partially restore the physical, biological, and chemical integrity of the Grand Canyon ecosystem.

to positioning in terms of price, service, and submarket, as well as to geographic location at national, local, and site-specific scales (see Figure 6.2). As retailing has become ever more competitive, optimizing the location of retail outlets has received increased attention from private-sector decision makers.

Locating retail and other consumer-oriented facilities is fundamentally a geographic problem, a problem that has been addressed by geographers in both theory and practice. During the 1960s, geographer David Huff adapted spatial interaction models for use in retail location decision making (Huff, 1963). His models have been widely used by retail planners to estimate probabilities of patronage based on the attractiveness (e.g., size, selection, price) of an outlet

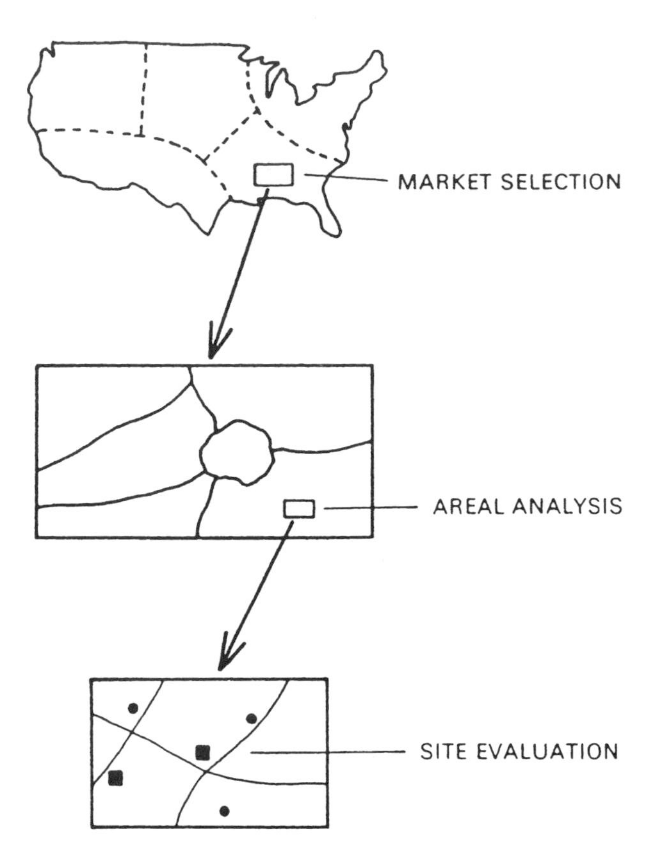

FIGURE 6.2 The success of a retail outlet depends to a large degree on its geographic location at national, regional, and local scales. All three scales must be considered in locational decision making. Source: Ghosh and McLafferty (1987).

and its location relative to possible markets and existing competitors. These models depict individual behavior as probabilistic, and they depict trade areas as overlapping and competitive rather than discrete and monopolistic (Haynes and Fotheringham, 1984). The models also have great flexibility, allowing the analyst to set the number of locations or criteria such as maximum distance to consumers or maximum capital costs.

Location-allocation models allow decision makers to optimize the location of all kinds of consumer-oriented facilities, public and private: stores, shopping centers, health care facilities, and schools. As retailing has become increasingly dominated by chains, the use of these models has become commonplace (Ghosh and McLafferty, 1987).

Dispute Resolution in the Courts

If it is true that geographic perspectives are important in developing policies, it is also true that such perspectives are important in defending or challenging such policies in court. Geographers frequently serve as expert witnesses in court cases where the legality of policies is tested. They have addressed residential segregation in cities for the U.S. Commission on Civil Rights; dealt with federal land management and environmental damage on Indian reservations in Indian Claims Court; devised busing plans for integration of school systems; and examined the human implications of physical processes such as accelerated erosion, river channel changes, flooding, coastal change, and lake dynamics in a variety of courts and administrative law proceedings (e.g., see Sidebar 6.3).

As expert witnesses, geographers engage in three types of activities. First, they provide a context for dispute resolution by defining the geographic characteristics of the area in dispute, the area's functional characteristics, and the systems that operate within it. Second, they testify about disputed facts. In cases dealing with housing discrimination, for example, the alleged discrimination must be defined accurately, demonstrated statistically, and mapped precisely. Third, geographers are frequently asked to apportion responsibility in court cases by defining what outcomes resulted from the actions of the defendant and what are likely normal outcomes on the social or environmental system. Accelerated erosion on valuable agricultural lands, for example, may be a product of negligent land management but also a product of entirely natural processes.

NATIONAL DECISIONS

Geographers participate in national-scale decisions by illuminating the dynamics of regions and the flows of materials, information, and people. A particularly strong field of contribution has been in transportation systems and policy. In addition, geographers help society understand its physical environment, contributing to our understanding of environmental impacts and resource use. At this scale, in both public and private organizations, the contributions of geographers are often cloaked by confidentiality and proprietary rights. The following sections illustrate some key contributions.

Energy Policy

A vivid example of the application of geography's perspectives and tools in national policy making comes from the energy "crises" of the 1970s. In 1979 the U.S. Department of Energy was developing a plan to allocate scarce gasoline in the event of an oil import disruption, and the states were urgently concerned about equitable treatment. David Greene, a geographer at Oak Ridge National Laboratory, developed an analytical model for understanding the determinants

SIDEBAR 6.3 Geography and Dispute Resolution: Electoral Redistricting in Los Angeles County

In 1988 the Mexican American Legal Defense Fund brought a suit against Los Angeles County challenging the 1981 reapportionment of supervisorial districts. The suit claimed that the boundaries of the five supervisorial districts established by reapportionment (see Figure 6.3) were drawn to fragment the Hispanic community. The Hispanic Opportunity District (shaded area in Figure 6.3) was presented by the plaintiffs as an example of a district in which a Hispanic supervisor could be elected.

A demographic analysis of Los Angeles County by geographer William Clark and demographer Peter Morrison (Clark and Morrison, 1991; Morrison and Clark, 1992) provided the context for adjudication of this dispute. Their analysis revealed that, at the time of the 1980 census, Hispanics made up 27.6 percent of Los Angeles County residents but only 14.6 percent of all voting-age citizens. If residents and citizens were similarly dispersed within the county, this disparity would be largely immaterial. In Los Angeles County, however, Hispanic noncitizens are more concentrated in certain parts of the county. Consequently, it is possible to form majority-Hispanic electoral districts that encompass one-fifth of all residents in the county (to achieve representational equality) but not nearly one-fifth of those who are entitled to vote (which determines electoral equality).

Clark and Morrison showed that the Hispanic Opportunity District (Figure 6.3) comprised 46.3 percent of Hispanic residents and 20 percent of all residents in Los Angeles County but only 14.4 percent of the voting-age citizens—well below the required 20 percent to qualify as an electoral district. They suggested that this apparent paradox was the result of several geographically driven demographic processes. First, there is a difference in the age structure of the Hispanic and non-Hispanic populations; only 61 percent of the county's Hispanics are 18 years or older compared with 77 percent of non-Hispanics. Second, some central areas of the county are heavily populated by adult Hispanic immigrants who are not citizens. Third, Hispanic citizens are more dispersed in the county than noncitizens.

Although the district court ruled that the 1981 reapportionment did not intentionally dilute voting strength, it did have the effect of preventing Hispanics from attaining a majority in a single district.

of gasoline use for highway transportation. His analysis was the first to recognize differences between the states in terms of per-person or per-household consumption of gasoline, based on such factors as differences in population distributions and associated differences in trip length for the journey to work, shopping, and social interactions. During negotiations with the states, the model was used successfully by the White House to resolve issues of equity.

In this same period an analytical modeling system, conceived in large part by geographer T.R. Lakshmanan, was one of the nation's major tools for forecasting the environmental consequences of different energy policy options. The model was useful because it addressed such highly geographic questions as the possible effects of coal development in the western United States on emissions from smokestacks in the east.

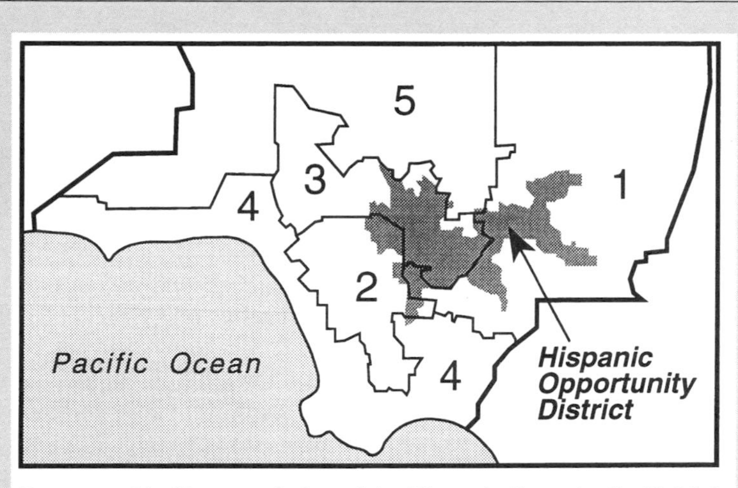

Demographic Characteristics of the Hispanic Opportunity District

Characteristic	*Apportionment Base* Total Population	Voting-age Citizen Population
Number of residents	1,495,466	553,105
Number of Hispanic residents	1,003,236	226,791
HOD's percentage of all county		
Residents	20.0	14.4
Hispanic residents	48.6	42.9
Percentage of population that is Hispanic	67.1	41.0(46.3)*

Source: 1980 Census of Population, including unpublished portions.
*The 46.3% figure refers to self-reported citizenship, uncorrected for misreporting.

FIGURE 6.3 Map of the Los Angeles County area showing the supervisorial districts established by the 1981 reapportionment (numbered polygons) and the Hispanic Opportunity District (shaded area). Source: Clark and Morrison (1991, p. 715).

Economic Restructuring and Competitiveness

Discussions of economic policies at a national scale often follow one of two related lines: concerns about a nation's standing with respect to other nations in terms of productivity, per capita income, or broader measures of living standards, or concerns about economic trends within the nation. Both of these concerns are integral to national economic "competitiveness," defined as the ability to combine rising living standards with increasing flows of trade and investment.

Many geographers have argued that viewing competitiveness in terms of national interest and national-scale policy making is overly narrow and can lead to bad decisions. They have instead tended to focus on the interplay of global and local processes. For example, one concern has been with the impact of public and private multinational institutions on local economies, especially local conflicts over labor issues that are related to adverse effects of a global competition in wage rates. A recent study by geographer David Angel for the U.S. Department of Commerce, for instance, has shown that alliances between multinational institutions and local groups are often associated with innovativeness and that such alliances strengthen local economic performance and benefit local economies (Angel, 1994).

Geographers generally consider competitiveness in terms of nation- or region-specific characteristics developed in response to place-specific economic histories. Economic development practice in many regions is moving away from pursuing large manufacturing facilities toward the redeployment of local assets, including a recognition of the potential of some service activities to foster economic growth, even in rural areas. This redeployment requires careful assessment of place-specific labor skills, product markets, technologies, and capital base, and an assessment of current and potential interactions with other regions or nations. Policy measures to improve competitiveness should be similarly place specific. In addition to the blunt tools of tax abatements or other subsidies, such policy measures can include targeted worker training, improved communications linkages, and arrangements for technical assistance (Glasmeier and Howland, 1995).

In these senses and others, geographers are involved by emphasizing the persistence of place distinctiveness and its effects on economic transformation. Much of the current research on industrial structure and place distinctiveness involves a great deal of labor-intensive personal observation, an extension of geography's tradition of fieldwork.

Technological Hazards

Building on a tradition of research on risks associated with natural hazards such as floods and droughts, geographers contribute both in theory and in practice to risk assessment for technological hazards. Many technological hazards, such as toxic waste disposal, chemical and nuclear accidents, and advanced weapons proliferation, are quite place specific. Consequently, evaluating the risk to human populations from such hazards calls for geographic perspectives (e.g., see Sidebar 6.4). Geographers have pioneered the development of risk assessment methodologies, hazard taxonomies, and hazard theories, and they have developed new understandings of how different groups perceive risk.

The solutions to many technological hazard problems are also strongly geographic. Over the past two decades, Congress passed an average of 23 laws per session dealing with technological hazards (Cutter, 1993), but many of these laws treat the nation monolithically and do not consider the "place matters" axiom of geography. Understanding and solving problems such as toxic waste disposal and high-level radioactive waste disposal call for the use of geography's core concepts (see Chapter 3): diffusion and dispersion; definitions of regions;

and spatial flows, past, present, and future (see Sidebar 6.5). For instance, airborne releases of contaminants from manufacturing facilities and from atmospheric tests of weapons are defined largely by the physical processes of atmospheric motion, but the health risks are related to the demographic characteristics of the

SIDEBAR 6.4 Geography and Emergency Response Planning for Nuclear Power Plants

Accidents at nuclear power plants can pose serious radiation hazards to surrounding populations if people and governments are poorly prepared—as at Windscale, England, in 1957 and Chernobyl, in the former Soviet Union, in 1986. The 1979 accident at Three Mile Island (TMI) in Harrisburg, Pennsylvania, prompted the federal government to require emergency plans for all nuclear power plants that would avert public exposure to radiation. The TMI Public Health Fund decided in 1985 to undertake the preparation of its own plan based on the state of the art that had emerged since the 1979 accident. Based on a national competition, the contract for this effort was awarded to a multidisciplinary national team of researchers led by geographer Roger Kasperson. The team included several other geographers as well—Dominic Golding, John Seley, John Sorenson, and Julian Wolpert.

The existing emergency plan for TMI had many serious shortcomings: objectives were vague and provided insufficient guidance, the capability to anticipate accident conditions and to initiate early precautionary responses was poorly developed, and the response plans were hindered by the rigidity of command-and-control operations and pro forma adherence to federal criteria. Response plans also were geographically rigid, conforming to the Federal Emergency Planning Zone of a 10-mile radius, even though radioactive materials could be dispersed beyond this zone. There was undue emphasis on evacuation, when such responses as sheltering in place might be more appropriate under certain circumstances.

The plan developed by this team (see Golding et al., 1992, 1994; see Figure 6.4) sought to remedy these shortcomings, offering an alternative and more ambitious approach to emergency preparedness. Although it was designed specifically for TMI, the plan has a conceptual basis that can be used at all nuclear power plants. Two principles guide the model plan. First, people are more likely to respond reasonably and effectively if they are well informed during an emergency, if they understand the information, if they understand why particular actions are necessary, and if well-considered and supporting resources are in place. Second, an effective plan should capitalize on the problem-solving abilities and special knowledge of persons and groups in emergency situations, thereby ensuring flexibility, resiliency, and efficiency in emergency response.

To overcome geographic rigidity the plan encourages flexible boundaries for emergency planning and preparedness and adopts three nested planning zones. The inner zone is within 5 miles of the plant, the area of highest risk that requires the most rapid response. Evacuation of the entire zone before core melt begins is recommended. The middle zone, about 5 to 25 miles from the plant, is an area of lower risk and flexible responses, ranging from sheltering to evacuation, depending on geographic conditions. The outer zone, typically beyond 25 miles, involves sheltering or evacuation/relocation should hot spots emerge.

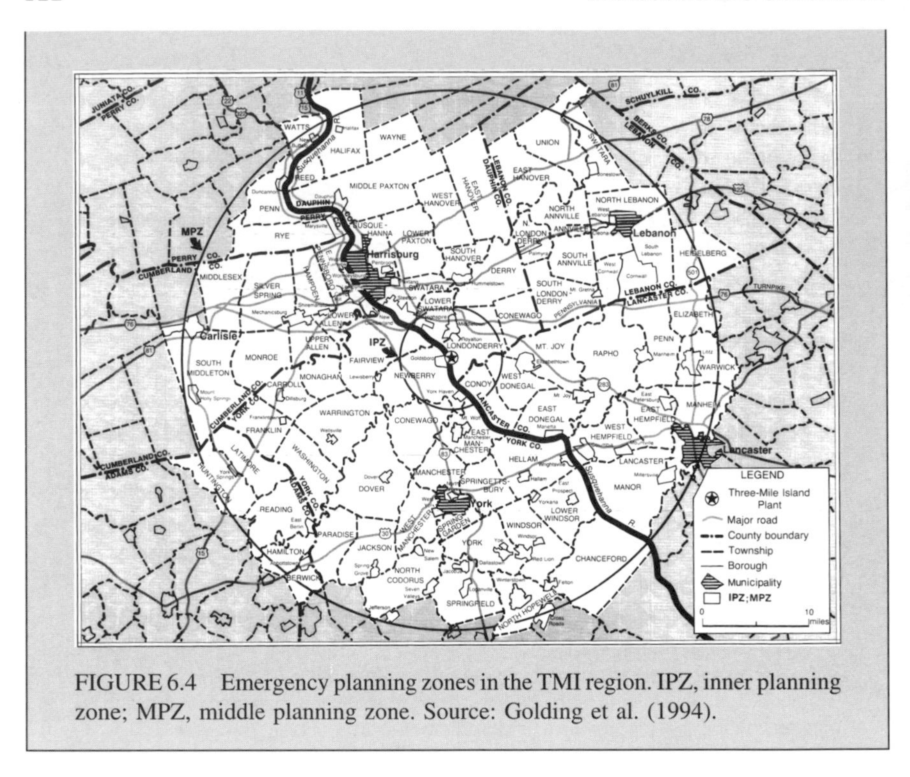

FIGURE 6.4 Emergency planning zones in the TMI region. IPZ, inner planning zone; MPZ, middle planning zone. Source: Golding et al. (1994).

population in those regions. Risk assessment of air releases therefore must account for human and physical geography in an integrated regional analysis.

An important aspect of the safe disposal of high-level nuclear waste is the design of safe transportation systems to connect waste generators with disposal sites. The problem is to choose the route that will expose the fewest people to risk while at the same time minimizing travel time and maximizing efficiency. Designing a transportation network that balances these competing considerations draws on the experience gained from solving similar problems associated with delivering other public services. Geographers are centrally involved in designing and operating such systems for radioactive and other hazardous waste transport in the United States.

National Floodplain Policy

The Great Flood of 1993 on the Missouri and Mississippi River systems resulted in more than $10 billion in property losses (see Sidebar 6.6). Recognizing the importance of wise investment of reconstruction funds and the need for adequate understanding of floodplain processes in the affected areas, President Clinton created the White House Interagency Floodplain Management Task Force

SIDEBAR 6.5 Geography and Nuclear Waste Disposal

A geographic perspective on long-term climatic change proved valuable during a recent Center for Nuclear Waste and Regulatory Analysis (CNWRA) assessment of possible future climate states at Yucca Mountain, Nevada, the candidate location for a permanent geologic repository for much of the nation's nuclear waste. The CNWRA convened an expert panel of five climatologists in 1993 (three of whom were geographers) to elicit their views about climate variability and change that may occur at Yucca Mountain over the next 10,000 years (DeWispelare et al., 1993). Climate changes that lead to wetter conditions at Yucca Mountain could affect the repository's ability to isolate radioactive wastes from the environment.

Panel members were asked to evaluate and integrate a diverse literature. Longer-term climatic trends, for instance, had to be assessed primarily from the paleoclimatic literature, with data sources ranging from analyses of pack-rat middens to changes caused by Milankovitch orbital/radiative forcing. Knowledge of the local and regional synoptic climatology and the weather station record informed the panel of the higher-frequency temporal and spatial variations. Climate model simulations of warmer future climates under increasing concentrations of atmospheric greenhouse gases also had to be considered.

After assessing and integrating this broad spectrum of information, all five climatologists agreed that a local warming of less than 3.5°C was plausible in the short term, followed by a longer period of cooling (see Figure 6.5a). The forecast cooling arose largely from an anticipated decline in extraterrestrial irradiance over the next 10,000+ years. Four of the five climatologists also thought that Yucca Mountain would likely remain dry well into the foreseeable future (Figure 6.5b).

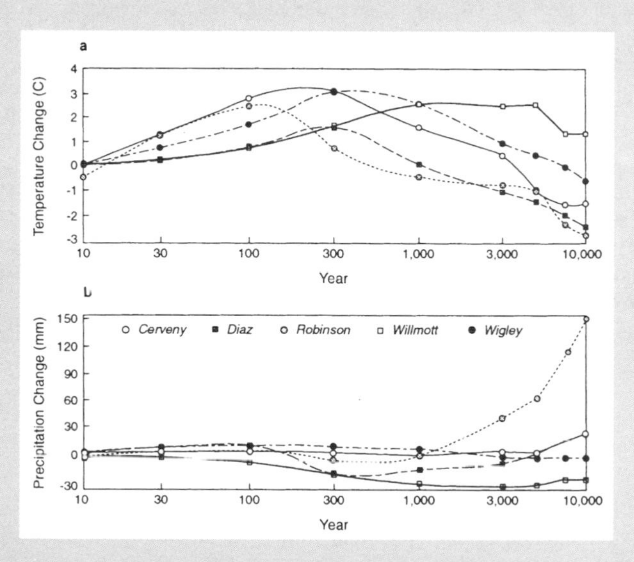

FIGURE 6.5 Predictions of five expert climatologists (each denoted by a different symbol) of (a) temperature and (b) precipitation trends during the next 10,000 years at Yucca Mountain, Nevada. Source: DeWispelare et al. (1993).

SIDEBAR 6.6 Mississippi Floods: How Big Is BIG?

Over a period of two decades, James C. Knox and his students at the University of Wisconsin, Madison, have investigated the physical evidence of changes in the regional river systems of the Upper Mississippi River system. Using sedimentological evidence, they have pieced together a comprehensive history of large flood events on the Upper Mississippi over the past 25,000 years. They used deposits of sediment left by these large floods to determine the depths of flow and then applied a standard engineering program model to calculate the likely amount of water in the various events.

The results of their work show that between 25,000 and 14,000 years ago the Laurentide Ice Sheet controlled the flow of water into the system, causing major aggradation by sand and gravel. By about 14,000 years ago, the ice sheet began to melt and retreat, creating numerous proglacial lakes dammed behind moraines. Discharges from these lakes, whether of low magnitudes on a regular basis or of occasional high magnitudes in catastrophic releases, were sediment poor and highly erosive, resulting in the net removal of the previously accumulated sediment. The largest outbreak flood, the Lake Agassiz flood, occurred sometime before 10,800 years ago and had a peak discharge of about 30,000 cubic meters per second. In the time period between that mega-flood and the present, changes have been less radical.

During the period between the Lake Agassiz flood and the present, very large floods have occurred, many of them of a magnitude similar to the great flood of 1993 on the Missouri and Mississippi River systems. When viewed on this long time scale, the 1993 flood does not appear to be particularly unusual. Without suggesting that human actions have had no effect on rainfall runoff and flooding, the policy implications of this finding are that (1) floods of the magnitude of the 1993 event should be expected to occur from time to time, and river managers should account for such events even with major flood control works in place; (2) floodplain areas are likely to be inundated despite the major levee systems now in place; (3) events of the magnitude of the 1993 event have occurred previously without the present human-induced changes in the system, such as altered land use on the uplands and draining of lowland riparian wetlands. Knox (1993) concludes that very large historical floods appear to be a result more of climatic factors than of land use. Perhaps regional residents are better off learning to live with the river and its floods rather than investing huge amounts of public capital in attempts to control a process that is climatically driven.

to make recommendations to the federal government for floodplain recovery and management. The task force is chaired by Brigadier General Gerald E. Galloway, a geographer at the U.S. Military Academy.

Data for the task force's work was prepared by the Scientific Assessment and Strategy Team (SAST). The team, which was directed by geographer John Kelmelis, spent several months at the Sioux Falls, South Dakota, data center of the U.S. Geological Survey (USGS) creating a database of useful information for the task force. The database included information on hydrology, geology, landforms, soils, structures (levees, dams, dikes), land use, hazardous and toxic materials, threatened and endangered species, and wetlands. Data were organized and presented as products from a geographic information system, which is capable of analyzing associations among the various types of data and provides the ability to map in detail individual reaches of the river system and adjacent environments (see Plate 10).

The work of this task force continues a tradition of contributions by geographers to federal management of flood problems. Previous reports by Edward A. Ackerman for the President's Water Resources Policy Commission (1950) and by Gilbert White for the Bureau of the Budget Task Force on Federal Flood Control (1966) have strongly influenced national floodplain management policies over the past five decades.

National Information Infrastructure

Government and business requirements for digital spatial data have stimulated a major effort to develop a national spatial data infrastructure (NSDI) along with associated spatial data transfer standards (see Chapter 5) and a national geospatial data clearinghouse (e.g., NRC, 1993b). Spatial data are useful in a wide variety of commercial applications ranging from marketing to navigation. Such applications require a publicly accessible spatial data infrastructure, digital cartographic databases, and global positioning systems (GPSs). Spatial data and associated geographic information analysis methods are also emerging as an integral part of national health policy discussions because many managed health care programs serve populations of geographically defined areas (e.g., Lasker et al., 1995).

Development of the NSDI has resulted in the creation of national georeferenced databases containing information about individuals. Such databases have fueled explosive growth in applications such as political canvassing, advertising, and travel. Geographers are contributing to the development of databases that allow easy access to such information. They are also engaged in research to assess who should have access to the data, how such access will affect the distribution of information and power in society, and what new legal and ethical principles will be necessary to control abuse in this rapidly developing area (Pickles, 1995a).

Creation of the NSDI is central to providing spatial data to the nation. In April 1994 President Clinton signed an executive order supporting the NSDI. The project will be directed by the Federal Geographic Data Committee through the Cartographic Requirements, Coordinating and Standards Program of the USGS. The NSDI will lead to the establishment of a geospatial data clearinghouse—a distributed network of geospatial data producers, managers, and users linked electronically. The clearinghouse will publicize data availability, facilitate links between data suppliers and users, and ultimately provide direct access to data.

Geographers are also involved in several other research initiatives that will contribute to the NSDI, including the following:

- The National Center for Geographic Information and Analysis leads research into the fundamental design and operations of GISs, as well as related efforts (e.g., the Alexandria Project) to develop methodologies and tools for the creation of a digital spatial "library." This library will allow decision makers, researchers, and the general public to easily access georeferenced information through the Internet.
- The U.S. Environmental Protection Agency's (EPA) Environmental Monitoring and Assessment Program (EMAP) monitors ecosystem changes at various spatial scales. The hierarchical database structure developed for EMAP now serves as a basis for much of EPA's environmental monitoring and analysis activities.
- The USGS's National Digital Cartographic Database supports the production of paper and electronic maps (see Sidebar 6.7).
- The U.S. Bureau of the Census has developed the Topologically Integrated Geographical Encoding and Referencing (TIGER) database (see Chapter 4).
- The U.S. Department of Energy's Environmental Spatial Analysis Tool (ESAT) project evaluates waste management alternatives.
- The National Center for Health Statistics has developed a cancer atlas.
- The U.S. Department of Transportation's Bureau of Transportation Statistics has undertaken a project to compile, analyze, and distribute data about the nation's transportation system.

SIDEBAR 6.7 Geography and the NSDI: USGS Geographic and Spatial Systems Program

In addition to playing a major role in developing standards for digital geographic and cartographic data and coordinating the NSDI project, USGS researchers have been exploring ways to integrate and apply spatial data in policy and management contexts. The USGS's Geographic and Spatial Systems Program "supports research to develop and test new and innovative theories and techniques to manage spatial data, including research in new techniques for modeling, analyzing, and visualizing spatial data in GIS and in automated cartography, image processing, and land characterization" (Kelmelis et al., 1993, p. 36). This research underpins national policies dealing with human-environmental interactions. A significant component of these policies involves land surface characterization. As part of a research program directed toward integrating GIS technology, remote sensing technology, and environmental simulation models, a new 159-class dataset of land surface characteristics has been constructed for the conterminous United States (see Plate 11). The digital representation of this dataset is used in the land process components of general atmospheric circulation, watershed, hydrologic, ecosystem, and biogeochemical cycle models.

Geographers participating in the above efforts design and develop data structures, methods for data manipulation and display, and tools for interactive geographic information analysis and visual display.

One of the greatest obstacles confronting research and development efforts, now and in the foreseeable future, is a dearth of people educated in spatial analysis, visualization techniques, and applications. Collecting and storing data are only part of the process; understanding and communicating its significance also are required. Thus, the development of the NSDI to support policy making challenges not only our technological expertise but also our education system.

INTERNATIONAL DECISIONS

The variability and connectivity of places are paramount for understanding the modern phenomenon of ''globalization,'' be it the international economy or environmental change. This realization has pushed geographic approaches and geographers to the forefront of many emerging global issues with both science and policy relevance. In this role, geographers are engaged in several interdisciplinary and integrative collaborations that bridge the natural and social sciences. This is particularly true in the efforts by geographers in the International Council of Scientific Unions to marshall scientific expertise in studies of the environment and global change, as noted in the examples below.

Responding to Global Environmental Change

From Stockholm 1972 to UNCED-Rio 1992, the international community has been acutely concerned about changes being imposed by humankind on the biosphere and the consequences of these changes for such important issues as sustainable development, biodiversity, and climate change. Both researchers and policy makers understand that such changes are anchored in human-environment relationships, are created by complex human activities, and have important consequences that vary significantly by locale (see Sidebar 6.8).

If the world becomes warmer, for example, some areas will be drier while others will be cooler and wetter (Henderson-Sellers, 1995). Where such changes take place matters, particularly in regard to the local economies and abilities of societies to adjust. Most assessments concur that agriculture in the tropical areas of the world may suffer the most from warming, exacerbating already serious food problems. In contrast, agricultural impacts in the developed world, while important for individual farmers and local areas, are embedded in systems that provide far more latitude for adjustments (Parry, 1990; Appendini and Liverman, 1994).

Geography remains central in setting the agenda for scientific research to address the complexity and integrativeness of human and social behavior related to global environmental change. Geographers have helped to launch international

SIDEBAR 6.8 Geography and Adaptation to Climatic Change in Major River Basins

One of the concerns associated with potential global environmental change is that less developed countries will be hardest hit because they will be least able to adapt. The latest assessment of global warming by the National Research Council Panel on Policy Implications of Greenhouse Warming (NRC, 1992b) was relatively optimistic about the ability of the United States to cope with anticipated changes. However, the panel was concerned that the less developed regions of the world would have greater difficulty in adapting to change because of the cumulative effects of population growth, weak infrastructures, and migration. One attempt to address this issue was led by geographers William Reibsame, James Wescoat, Gary Gaile, and Richard Perritt, with support from the EPA. Their study, which is described in Strezepek and Smith (1979), involved more than 20 scientists and water resource managers who assessed the potential effects of possible adaptive responses to climate change in five of the world's major river basins in less developed countries: the Mekong, Zambezi, Uruguay, Indus, and Nile river basins.

The researchers linked the environmental consequences of possible global warming to hydrology and social conditions in each basin by extracting regional climate change scenarios from atmospheric circulation models and feeding the results into regional simulations of basin hydrology and water resource management. Regional responses included changes in flooding, hydropower production, and other natural resources. Local water resource managers in the five basins were asked how they would cope with the suggested scenarios, so that the project defined a realistic potential range of adjustments that policy makers might consider.

Many of the river basins were quite adaptable, although the specific adaptations were varied. Adjustments such as the construction of small hydroelectric plants (as opposed to large dams), with flexible irrigation practices for the Uruguay and Mekong river basins, and planned development of new dams on the Zambezi and Mekong rivers provided increased resilience to change. On the other hand, the single large reservoir serving the Nile basin could fail to meet Egypt's irrigation needs under several climate change scenarios, and the extremely complex irrigation systems of the Indus River basin, which already suffers from high water tables, salinization, and water distribution problems, could begin to return fewer benefits as climate changes. Although the project did not recommend specific policies, it did provide a reasonable framework for considering policy responses to possible changes in the near future.

collaborative research on greenhouse gases and climate change through SCOPE (the Scientific Committee on Problems of the Environment; Bolin et al., 1986). They are active in the development of reports of the IPCC (the Intergovernmental Panel on Climate Change) and of research agendas for various international global change science programs, including the International Geosphere-Biosphere Programme (IGBP) and the International Human Dimensions of Global Environmental Change Programme (IHDP). Geographic approaches and analytical techniques are central to such IGBP and IHDP projects as Data and Information

Systems (Townshend, 1992) and Land-Use/Cover Change (Turner et al., 1995), and geographers are also deeply involved with the Global Change System for Analysis, Research and Training and the Inter-American Institute.

In addition, geographers associated with the U.S. Agency for International Development (USAID) have been active in applying geographic perspectives and tools (particularly GISs) to strategies for coping with the relocation of refugees and the routing of supplies to support them.

Some of this work should find application in business strategy development, as global market conditions shift in the direction of environmental sustainability. For products and services to be competitive in national and regional markets a generation from now, business strategists must address the realities of "greener" policies on a global basis. Anticipating these changes in all their geographic complexity is becoming an increasingly important aspect of corporate planning.

Global Economic and Political Restructuring

During the past decade, the international economic and political order has undergone a profound transition. Formerly monolithic entities such as the Soviet Union and the Warsaw Pact nations have splintered into a host of diverse economic and political entities. Relationships among economies have changed as manufacturing has moved to developing countries with lower labor costs. Trade barriers have fallen as states in some parts of the world have sought closer economic ties with one another. New transnational networks have emerged as individuals and corporations have forged links outside traditional state structures.

Understanding global economic and political transformations is a complex task that is not the province of any one discipline, but geography has an important role to play. Indeed, one of the principal catalysts for the recent rediscovery of geography was concern over the inability of many U.S. citizens even to identify where particular changes were taking place, much less understand how they might be altering the political, economic, and environmental character of the planet (see Chapter 1). Geographic work on global political and economic restructuring provides insights into problems faced by decision makers by elucidating both the changing spatial organization of human and environmental phenomena and the effects of global changes on specific places (e.g., Johnston et al., 1995), as well as the relationships between them. At a regional scale, studies by geographers have sensitized policy makers to the problems and prospects of regional integration initiatives in regions such as the Middle East and Pacific Asia (Drysdale and Blake, 1985; Murphy, 1995a). At the local scale, geographers have helped public and private decision makers recognize the importance of geographic context when addressing resource management issues (see Sidebars 6.9 and 6.10).

The relevance of geographic thinking for efforts to understand global economic and political restructuring can be illustrated by considering three contemporary transformations of great significance: the growing reliance on market mecha-

SIDEBAR 6.9 Geography and the Development of Agricultural Policy

Societies throughout the world are struggling to develop efficient and effective approaches to the management of agricultural land. In many cases, small landholders are on the front lines of such struggles. Research has shown that the ability of small landholders to increase production and conserve natural resources depends on their knowledge base and the extent to which they are able to play a meaningful role in resource management decisions. Consequently, a growing number of resource management initiatives are focusing on the situation of small landholders.

One important example is a project led by Gerald Karaska, in which geographers are working with development anthropologists, economists, and others to assist the Sri Lankan government in its efforts to develop a program to increase the efficiency of water use and improve the lot of small holders without unduly sacrificing agricultural production and environmental quality. The project is part of a USAID initiative, the Systems Approach to Regional Income and Sustainable Resource Assistance (SARSA), which seeks to assist governments around the world in efforts to improve resource use. The project is designed to increase the control of resource users in management decisions through partnerships based on formal agreements between the state and small landholders.

Since its inception in 1993, geographers have played a leading role in the Sri Lankan project, and its structure bears the imprint of their perspectives. The watershed is the basic unit of analysis, and studies of spatial relationships between economic and environmental variables have been used to understand the complexities of water management issues. Moreover, intensive participatory interaction among resource users and state and local officials has been employed to develop site-specific resource management strategies. The project identifies subregions sharing particular human and environmental characteristics, and these are then used as the spatial frameworks within which planning decisions are made.

The project has already produced striking results, with farmers' organizations agreeing to share the flow of water and change their land uses according to their locations within watersheds. These results are expected to lead to national policy changes as well as to new laws that give farmers long-term usufruct rights to land and forest improvements.

Another example shows how geographers can link their academic interests in development with the "hands-on" development work of nongovernmental organizations (NGOs) to influence agricultural policies. Anthony Bebbington, formerly of the Overseas Development Institute and now of the International Institute for Environment and Development, bridged the academic-policy gap in Ecuador and associated Andean regions (Bebbington, 1994). Working with the Inter-American Foundation and the Fundac para el Desarrollo Agropecuario (FUNDAGRO) in Ecuador, Bebbington expanded the meaning of farming systems to include indigenous farm organizations and investigated their role within the larger environmental system and the regional political economy. Special attention was given to federations of *campesino* communities, NGOs, the modern church, and certain agencies of the state and how these entities were used by the *campesino* to negotiate relationships with the market and the state. Bebbington's ethnographic work showed how agrarian societies interact with the NGOs, and he demonstrated diverse ways in which households, communities, and organizations interact and respond to changing economic and environmental conditions.

Bebbington's work, in association with FUNDAGRO, made NGOs rethink their relationship with various indigenous farmers, particularly the *campesino* federations. The success in working with these federations has fed into international and national agricultural research institutions in Ecuador and the Andes, stimulating policy that involves such federations in determining agrarian policy and the nature of Andean agriculture. Andean agricultural development now incorporates indigenous farmer organizations and grassroots NGOs.

SIDEBAR 6.10 Thornthwaite and Agricultural Practice

In addition to his seminal work on potential evapotranspiration and the climatic water budget (Mather and Sanderson, 1996), C.W. Thornthwaite was a pioneer in applied climatology, especially in relation to commercial agriculture. In a 1956 article that highlighted the applications of operations research, *Fortune* magazine devoted nearly a page to Thornthwaite's achievements "down on the corporate farm."

The story illustrated how he was able to organize the planting and harvesting of peas and later of other crops grown on some 50,000 acres controlled by Seabrook Farms. Thornthwaite's "harvest design" revolutionized the produce industry, making it possible to efficiently schedule the whole farm program from planting, to irrigation, to harvest of vegetables, without overlap of vegetables or glut of produce waiting in the factory yard at processing time. According to *Fortune*, it also eliminated night crews, stabilized the work force, and improved labor relations, from which Seabrook Farms realized a substantial savings.

Thornthwaite's own account of this use of an operations research approach appeared in the *Journal of the Operations Research Society of America* in February 1953 (Thornthwaite, 1953).

nisms, the rise of democratic movements in various parts of the world, and the changing nature of the world economy. In the first case, increased reliance on the market shifts the relative influence over capital flows and accumulation from the public sector to the private sector. Geographers studying this process have highlighted the integration of financial markets, and the corresponding accelerated speed of capital circulation, a process dubbed ''the annihilation of space by time'' by Harvey (1989). While places are able to mitigate the threat of capital mobility, at least for a time, taking advantage of economic niches and centers of power within the ''space of flows'' (Castells, 1989), this acceleration increases the economic uncertainty faced by all places and the complexity of the ways in which growth and decline in different places, and at different spatial scales, are interrelated (Thrift, 1989). As a consequence, the interests and economic prospects of distant regions are continually bound together in new ways (Erickson and Hayward, 1991), and there is increasing demand for transportation and communications services (Warf and Cox, 1993).

In the second case, studies of political changes such as democratization have focused heavily on social and economic arrangements in individual states. Yet as a recent report to the National Science Foundation (Murphy, 1995b) makes clear, a geographic perspective on democratization is vital because it relates such concerns to underlying territorial, spatial, and environmental circumstances. How does a country's position in relation to international political and economic arrangements affect the ability of nondemocratic regimes to resist democratization initiatives? What threats do subnational regional inequalities pose for democratic regimes? To what extent does environmental change affect the emergence and sustenance of democratic regimes? Geographic work on questions such as these provides decision makers with information about developing trends and with ideas about the circumstances under which democratic institutions are likely to flourish (e.g., Central Intelligence Agency, 1995).

In the third case, two processes increasingly characterize the changing world economy and will continue to impact the world economy in the next century. The process of globalization describes the growing integration of various parts of the world into a global economy (Dicken, 1992). The process of urbanization describes the growing concentration of the world's populations in cities. Some large cities have become key hubs for the control and coordination of global finances and of the global economy itself (Dieleman and Hamnett, 1994).

The growth of world cities is not without problems (Sassen, 1991). It is often in these cities that there are large gulfs between rich and poor, a polarization of social status that is exacerbated by increasing migration between developing and developed societies (Fainstein et al., 1992). The flows of population within North America and to the United States from nearby less successful economies is similar to flows in other parts of the world economy. These flows are having and will continue to have powerful effects on the urban structure and infrastructure of large and small cities in the United States. Policy debates over closed and open door immigration policies are only one manifestation of local problems that are the outcome of international changes.

Often independently of the growth of these command and control centers there is a growing urbanization of the world population as a whole. In just a few decades the number of very large cities has increased fourfold and many of these new large cities are in Africa and South America, where the infrastructure still lags behind the needs of the local populations. Solutions to these seemingly unsolvable problems are at the heart of creating a stable twenty-first century.

Technology, Service, and Information Transfer

The spread of ideas from place to place is a focus of geography, and potentially one of its most direct applications is to international trade. Here, geography's contributions could include helping decision makers understand the local context for trade with international partners and the importance of long-term social, economic, and environmental sustainability within this local context. International technology transfer and trade from the United States depends fundamentally on responding to local realities in other countries, which calls for certain kinds of expertise that geographers are expected to possess. Generally, however, geography as a discipline has been a minor player in this arena, partly because it has too seldom connected its best scientific insights with business decision making at the international scale and partly because its experts are so few.

Geographers also transfer their perspectives and techniques to more local problems. When children in a rural school district, for example, are bused to school, each child's journey is a regional interaction. Questions such as the optimal placement of schools or the location of attendance boundaries for an existing set of schools to minimize the cost of student movement are important. The location-allocation modeling methods discussed earlier in this chapter have been developed to solve such problems in the United States, and these spatial decision support systems are now routinely used to find the best locations for many social and economic infrastructure elements in other countries as well (see Sidebar 6.11).

Geographers contribute to information policy by applying their expertise on evolving communication networks used by business and government, which affect international strategy development and implementation. Most directly, of course, geography contributes by transferring its own information and tools such as GIS hardware and software. In some cases this involves development of hybrid GIS tools to take advantage of computers and software already in place at regional levels in developing countries (Yapa, 1989). In addition, geographers have played an increasingly important role in efforts to develop international standards for georeferenced spatial data. These contributions include leadership of a multiyear effort within the International Cartographic Association to develop scientific and technical characteristics by which any national or international spatial database transfer standard can be assessed (Moellering and Wortman, 1994; Moellering and Hogan, in press) and participation (as Project Leaders and Experts) in the International Standards Organization Technical Committee on Geographic Information/Geomatics (ISO/TC 211).

Hunger

In 1990 more than one-fifth of the world's population suffered from micronutrient deficiencies and nutrient-depleting illnesses. More than 700 million people faced chronic undernutrition. Tens of millions were highly vulnerable to famine. Hundreds of millions of children suffered from below-normal growth. These statistics illustrate the many faces of hunger and inadequate diets relative to the kind and quantity of food required for growth, activity, and the maintenance of good health.

The attack on hunger is a multidisciplinary and highly fragmented effort to which geographers contribute in a variety of ways (Bebbington and Carney, 1990). In collaboration with researchers in other disciplines, geographers are working to help policy makers understand the structure and root causes of hunger. They have identified three distinctive hunger situations: regional food shortages in which there is not enough food available in bounded areas; household food poverty in which there may be sufficient food available in an area but some households do not have the means to obtain it; and individual food deprivation, in which there may be adequate food, but food may be withheld from individuals, their special nutritional needs may not be met, or illness may prevent proper absorption of their diet.

SIDEBAR 6.11 Case Study: Using Geography to Locate Health Care Facilities in India

In 1972 the Central Ministry of Health in India implemented a program to make primary health care more accessible in rural areas by establishing one primary health care unit for approximately every 15,000 people. In the state of Karnataka, located in southern India, the directorate of health implemented this policy by defining small regions called "population blocks." Each of these blocks, which contained approximately 15,000 people, was defined by administrators who were reasonably familiar with the local territory. After approving this implementation plan in 1972, the state health planning unit approved the establishment of new primary health units only in "vacant population blocks" in which government health care services did not already exist.

In one district in Karnataka (the Bellary District, with a population of 1.1 million people in 1971), 15 of 67 population blocks in the district were without a primary health unit. Geographers investigated the effects of establishing new health units in the vacant blocks on accessibility to health care services. They used a location-allocation model (discussed earlier in this chapter) to identify, in sequence, the locations of 15 places that if added to the existing health centers would most reduce the distance to the closest place offering government-supported health care services. In the first case, all places in the district were considered by the algorithm to be eligible for new health units. No requirement about locating in vacant population blocks was imposed (see Figure 6.6, top curve). In the second case, new health units were restricted to vacant blocks as required by state policy (Figure 6.6, bottom curve).

Figure 6.6 shows that the most significant reductions in travel distances could be obtained if new health units were not restricted to vacant population blocks. That is, the approach of designating population blocks and using them in the decision-making process forced decision makers to make less than optimal decisions. This information led the ministry of health to change its methods for locating health facilities (Rushton, 1988).

Researchers recognize that these situations are linked as in a cascade, at the bottom of which are individuals who are hungry. In times of regional food shortages, a cascade of troubles plunge food-sufficient households into food poverty and adequately fed individuals into food deprivation (see Figure 6.7). This understanding has led to major improvements in the conceptual basis for famine early warning systems.

Geographers are also leading efforts by NGOs to alleviate global hunger. Geographers Robert Kates and Akin Mabogunje cochair the organization Overcoming Hunger in the 1990s, which has proposed four achievable goals for reducing hunger by the turn of the century: (1) eliminate deaths from famine, (2) end hunger in half of the poorest households, (3) cut malnutrition in half for mothers and small children, and (4) eradicate iodine and vitamin A deficiencies. These goals have become part of the mainstream global hunger agenda; they appear in the resolutions of the 159 nations participating in the 1992 International

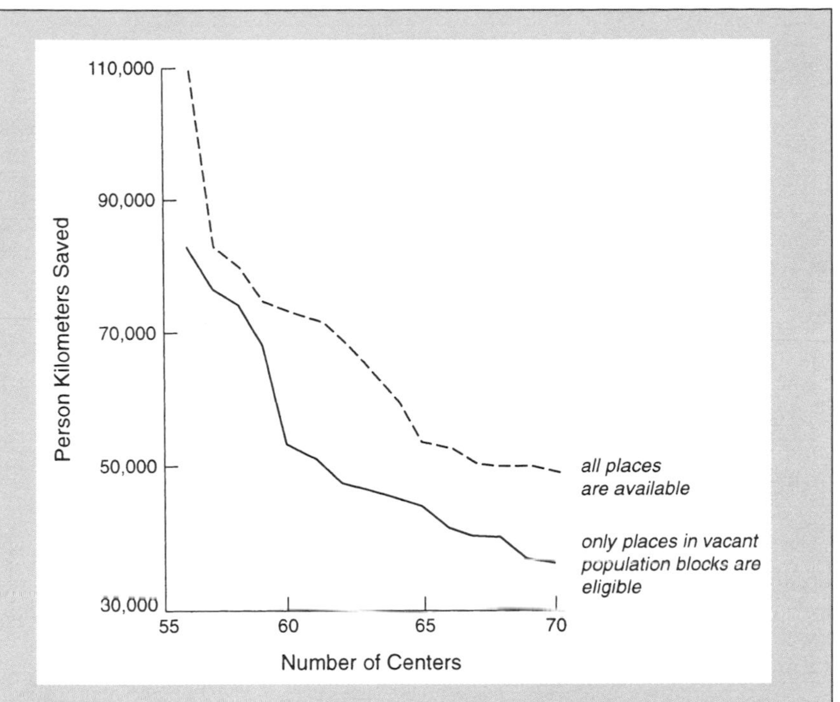

FIGURE 6.6 Calculated reductions in travel distances to health units in the Bellary District of India by optimally locating 15 new health clinics. *Top curve:* reductions in travel distances if no restrictions are placed on the locations of new health clinics. *Bottom curve:* reductions in travel distances if new health clinics are located in vacant population blocks. Source: Rushton (1988).

Conference on Nutrition and the 1993 World Bank Conference on Overcoming Global Hunger.

SUMMARY AND CONCLUSIONS

This chapter has illustrated a wide range of issues in which geographers contribute to decision making in both the public and the private sectors. The chapter has also illustrated areas of concern where geography is not now contributing in any significant way but could be given increased attention, resources, and access to policy makers. A common theme throughout much of this review is geography's role as a provider of distinctive perspectives, information, and technology for decisions. Geography provides a valuable way of thinking about human and environmental issues, a way of thinking that emphasizes connections in complex systems and the importance of locations and arrangements.

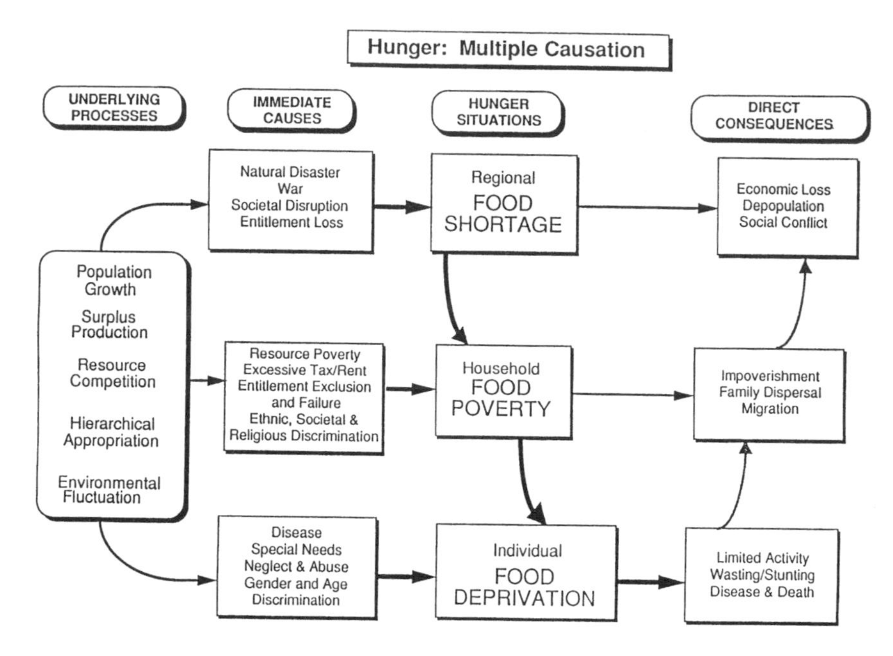

FIGURE 6.7 The causal structure of hunger.

Geography's contribution to decision making has a long and distinguished tradition starting with the founding of the United States. One of the earliest American geographies was Thomas Jefferson's *Notes on the State of Virginia,* a wide-ranging treatise dealing with a region extending from what are now the Middle Atlantic states into Ohio. Jefferson evaluated the development potential of the then-new United States through an analysis of its physical and human geography.

As the nation expanded westward, public- and private-sector policy makers relied on the analyses of exploration geographers. For example, George Perkins Marsh's *Man and Nature, Or, Physical Geography as Modified by Human Action* led directly to the creation of new federal resource management agencies in the late nineteenth century. John Wesley Powell helped define national reclamation policy. Henry Gannett was a leader in defining the emerging national forest system. Harlan Barrows participated in numerous New Deal national planning boards for river basins. Carl Sauer worked with the Soil Conservation Service during the New Deal era to improve land management, and C.W. Thornthwaite worked with the U.S. Department of Agriculture on climatic classifications for assessing moisture availability for crops. Gilbert White profoundly impacted the nation's geographic perspective on floods and flood hazard management through his research and his service on prominent committees such as the Hoover Commission Task Force on Natural Resources to the Bureau of the Budget Task Force on Federal Flood Policy.

In recent years, geography and geographers have been less prominent in some aspects of significant national decision making. For instance, the Wilderness System and Wild and Scenic Rivers System were enacted essentially without geographic considerations. Potentially significant and wide-ranging contributions by geography to foreign policy have been hampered by the small size of the Office of the Geographer of the U.S. Department of State. Highly geographic problems—ranging from the designation of formal wine appellations in California to U.S. policy for deciding whether to intervene in regional or national conflicts outside the United States—are sometimes not addressed by geographers at all. When they are, geographers often do not interact with decision makers. In this respect the discipline has fallen short of its potential to make meaningful contributions to American society (e.g., Berry, 1994).

The limited ability of the discipline to respond to the geographic needs of decision makers is a result of at least two factors: a shortage of formal connections between geographers and decision makers and the small size of the discipline. Consequently, the United States lags behind the rest of the developed world in its geographic sophistication in many aspects of decision making. To borrow a Cold War expression, there is a ''geography gap'' that may cost the nation dearly in terms of competitiveness and ability to achieve the twin goals of economic prosperity and environmental stability. Closing this gap depends on strengthening the foundations of the discipline along with its research and educational dimensions.

7

Strengthening Geography's Foundations

Geography and geographic approaches to problem solving have contributed significantly to the scientific and decision-making communities in the United States. This report was written, however, not only to explain these contributions but also to address what should be done to more fully realize geography's potential to address critical questions for science and society at large. The foregoing chapters have laid the foundation for this discussion by reviewing geography's perspectives, techniques, and contributions. This chapter now turns directly to addressing the committee's statement of task (see Chapter 1):

- to identify critical issues and constraints for the discipline of geography,
- to clarify priorities for teaching and research, and
- to link developments in geography as a science with national needs for geography education.[1]

Geography's ability to respond to the challenges inherent in its rediscovery is potentially hampered by several impediments. Despite three decades of growth in the number of professional geographers,[2] the geography community remains small relative to most other natural and social science disciplines. Few colleges

[1] The remaining two objectives of the assessment—to increase appreciation of geography in the scientific community and to communicate with the international scientific community about the future directions of the discipline in the United States—will be achieved by dissemination of this report.

[2] As noted in Chapter 1, the membership of the Association of American Geographers has increased from 2,000 to 7,000 since 1960.

and universities have large geography departments,[3] and geography is not taught at all at many institutions of higher learning,[4] including some of the nation's leading universities. Thus, many students—as well as faculty members and administrators—lack an appreciation for the importance of geographic perspectives.[5] Women and minorities are underrepresented in senior academic and professional positions relative to their numbers in the general population, and, at present, few minorities are entering the field.[6] This small human and programmatic base will make it difficult for the discipline to respond effectively to increased demands for attention—demands that are likely to increase still further in the years ahead.

Demands for increased attention are likely to come from several quarters. For example, efforts to infuse geography into the curriculum of the nation's schools will likely lead to increased demands for college-level teacher training in geography. Indeed, a surge in student interest in geography is already evident at the college and university levels (e.g., see Figure 1.1 and Appendix A). The use of geography by the research and decision-making communities (see Chapters 5 and 6) is likely to translate into a growing demand for college graduates with geographic training.

Realizing geography's potential requires more than addressing the problems presented by the discipline's small size and limited diversity, however. In several critical areas, geography's intellectual foundations need to be strengthened to ensure that its contributions to science and society are solidly grounded. Moreover, geographers need to work to overcome the view that geography is simply a descriptive subject with little analytical or technical depth. This is particularly critical given the growing demand for technical expertise on the part of geographers entering the labor market. At least as important, the appreciation and use of geography by nongeographers need to be fostered, so that the capacity to make use of the discipline's perspectives, knowledge, and techniques grows along with the capacity of the discipline to supply them. This includes enhancing the

---

[3]Of 27 major departments of geography in the United States selected for inclusion in the Gourman report (1993), 5 had between 20 and 25 full-time equivalent (FTE) faculty members, 12 had between 15 and 20 FTE faculty members, and the remaining departments had between 10 and 15. The 36 Ph.D.-degree-granting departments listed in the National Research Council's latest survey of research doctorate programs (NRC, 1995) averaged 15 faculty members and 33 doctoral students.

[4]Of the roughly 2,200 accredited four-year colleges and universities in the United States, about 250 offer undergraduate or graduate degrees in geography.

[5]This situation is extraordinary by world standards because geography is considered a core subject in most universities in Europe and East Asia.

[6]Support for this statement comes from a variety of sources. At the end of 1994, for example, only 1.4 percent of the members of the Association of American Geographers were African American, 1.5 percent were Hispanic, and 0.6 percent were Native American (AAG, 1995). In 1993 females and minorities received about 28 percent and 4 percent, respectively, of the doctorate degrees from the 36 Ph.D.-granting departments surveyed by the National Research Council (NRC, 1995). In 1990 women held only 5 percent of the tenured faculty positions in geography (Lee, 1990).

geographic competency of the general population and fostering better geographic training in colleges and universities.

STRENGTHENING GEOGRAPHIC RESEARCH IN SELECTED AREAS

If geography is to respond effectively to increased demands for its research expertise, the discipline needs to strengthen its intellectual foundations in areas that exploit its distinctive insights (see Chapter 3) while at the same time recognizing and meeting its responsibilities in certain technical and educational endeavors. To equip the discipline to meet its larger responsibilities to science and society that have been touched on throughout this report, attention must be focused on the following six research challenges:

1. disequilibrium and dynamics in complex systems,
2. an expanded concept of global change,
3. the local-global continuum,
4. comparative studies using longitudinal data,
5. effects of geographic technology on decision making, and
6. geographic learning.

The first four of these challenges involve research topics that exploit the integrative and synthetic traditions of the discipline and strengthen the discipline's connection to the larger family of science. These challenges share at least two attributes, being both complex and integrative. As such, they can be addressed only through the synthesis of a wide range of information and expertise. A major need of the discipline is to establish and maintain a balance between its synthetic tradition and its more recent movement toward specialization, and these four initiatives are designed with that balance in mind.

The last two challenges are important because, irrespective of disciplinary emphases, demands for the use of geographic technology in decision making and for teaching geography are growing. If geographers do not respond, there is a high likelihood that primary and secondary schools will continue to employ a simplistic, fact-based approach to geography education. There is also a risk that those employing geographic technology [such as geographic information systems (GISs)] in the public and private sectors will use that technology in inappropriate or inefficient ways.

Disequilibrium and Dynamics in Complex Systems

Most geographic models have been based on an assumption that system dynamics are dominated by strong and stable equilibria. Thus, numerical models assume that steady-state representations are adequate, and potentially nonlinear or chaotic behavior is tacitly suppressed by adjusting time/space parameters to be consistent with empirical observations and theoretical constructs. For quite some time, however, it has been recognized in principle that such simplifying assumptions are unrealistic.

Many fields of research would benefit if geographers were to incorporate nonlinear and complex dynamical phenomena more effectively in their models and other representations. It is well known in economic geography, for example, that under certain conditions spatial disequilibrium occurs. It is also well known that rates of interactions between places change as a result of endogenous changes in the geographic distribution of economic activities. In physical geography, evidence of nonlinear feedbacks or chaotic behavior has long been suspected in climate, stream-network, ecosystem, and landscape systems. Incorporating realistic disequilibrium and complexity into analytical and numerical models is a challenge that mathematically oriented physical and human geographers must embrace if their models are to better represent real systems.

A focus on complex nonlinear systems is important at least partly because it opens up the possibility of incorporating into theory the kinds of evolutionary path dependencies, contingencies, and irreversibilities that are stressed in much qualitative work on geographic and historical processes. Research on complex systems also will place geography centrally within an emerging scientific synthesis that embraces such disciplines as physics, atmospheric science, economics, computer science, and genetics and that treats complex dynamics as an interdisciplinary research frontier.

Expanding geographic research capabilities related to complex systems will require the training of early-career geographers in the necessary analytical and computer skills. Such training poses significant challenges to the discipline because the number of researchers interested in and equipped to work on this problem still is extremely small. Expanding geography's capabilities will require access to adequate computer resources, both in central computing systems and on the desktop, to undertake large numerical simulations that are becoming the primary method for investigating complex systems. Access to advanced visualization algorithms and media will be essential as well, as will research to develop more effective visualization methods.

Broadening the Concept of Global Change

Global change research has become a major interdisciplinary focus for science, comparable in breadth, organization, and required resources to such large comprehensive international scientific undertakings as the human genome project. Part of the appeal of global change research rests on the widespread perception that major adjustments in the Earth system are under way that are wholly or partially human induced and that need to be controlled or at least understood. To date, research on global change has focused primarily on the climate system. Yet equally important global changes are occurring in other facets of the physical environment and in social systems, including economies, populations, governments, and cultures. For instance, global climate changes are likely to be reflected in the distribution, quantity, and quality of water resources. Many societies have invested immense amounts of capital in structures to exploit water resources for economic development, and the functions of these structures may have to be altered to reflect new conditions that are different from those for which the structures were designed.

A broadened concept of global change would exploit at global scales geography's expertise in studying physical and social processes and their relationships. Such a broadened concept recognizes that climate changes do not operate in isolation from other environmental and social changes. Rather, they need to be studied together with other changes induced by human activities in order to be understood completely. Such changes include the global accumulation of pollutants in the biosphere—acid deposition, heavy metals in soils, chemicals in groundwater—as well as global biotic changes, specifically deforestation in tropical and mountain lands, desertification in dry lands, and species extinction, particularly in the tropics. Global social, political, and economic changes also need to be considered: the expected doubling of world population in as little as 50 years, the massive restructuring of the world economy, the changing role of the modern state, the flow of migrants across international boundaries, and the unpredictable and sometimes violent response of people to such changes.

In short, it is no longer sufficient to focus research efforts exclusively on specific climate changes or local case studies. Distinctive geographic approaches need to be developed to address the interplay of global and local changes (Meyer and Turner, 1994; Riebsame et al., 1994). Toward this end, studies might examine the interaction of global changes in particular places along the local-global continuum (see the next section) and in case-controlled comparative situations. Such efforts will require new types of collaborative initiatives, methods for case-study comparison, hypotheses of global interaction, and appreciation of different approaches to problem solving.

The Local-Global Continuum and Movement Across Scales

Global processes and events are increasingly connected; no matter where they occur along the local-global continuum, they have impacts at other places and other scales. The widespread recognition of this connectivity, commonly referred to as the micro-macro issue, has increased interdisciplinary concerns about how to link spatially variable events and processes, as well as the analytical problems involved in doing so. The challenge for geographers is to examine micro-, meso-, and macroscale mechanisms within a framework governed by such principles as the following: (1) causal mechanisms are best observed at local levels, (2) macroscale events are not always best explained by reducing them to local-scale events, and (3) macroscale processes do not always deterministically structure local-scale events.

Geography has a long tradition of grappling with scale problems, and its regional science component has devoted considerable attention to underlying analytical issues (e.g., Isard, 1975; Haynes and Fotheringham, 1984). Geographers recognize that answers to research questions are frequently scale dependent. Yet this principle and its ramifications for understanding phenomena and processes in a geographic scale continuum are not well recognized in the broader natural and social science communities, ecology being a notable exception. Why this is so reflects, in part, inadequate efforts by geographers and other spatial scientists to distill from their work the implications of scalar rules for major problems in science.

One impediment to understanding phenomena and processes in a geographic scale continuum is that they may be hierarchically nested in complex ways. For example, movements in the U.S. prime interest rate may trigger changes in international money markets, which in turn are filtered through various national economies into local economic decision making. The sheer complexity of these relationships strongly suggests that interactions across scale are not linear but involve thresholds and abrupt jumps between different conditions and outcomes and that outcomes vary considerably by locale and region.

Another impediment to understanding phenomena and processes in the scale continuum is the paucity of comparable georeferenced information at different scales. This problem has two components. The first is a problem of observation. There are simply not enough geographers—or scientists in any field—collecting data on local processes, and direct observations of nonlocal processes are rarely feasible. Remote sensing technology has had a dramatic impact on the ability to conduct analyses at multiple spatial and temporal scales, but not all data collection can be done by remote methods, as discussed in Chapter 4 (see Figure 7.1). Beyond observation, multiscale analysis poses fundamental problems of generalization. When data are collected independently at different scales of analysis, decisions must be made about the nature and scale of the data to be collected, and methods must exist to automatically transform that information so that it can be used at other scales.

A number of research communities are looking to geography for the scalar rules to guide multiscale analyses (IGBP, 1994; NRC, 1994; USGCRP, 1994). Many of the guiding principles logically rest within geography's intellectual domain, but the insights need to be sharpened, articulated, and presented in ways that are relevant to the larger scientific community. Despite evidence of a growing interest in scale within geography, much remains to be done in developing general conceptualizations of how processes at different scales affect one another and in developing quantitative and qualitative methods for identifying and analyzing such processes.

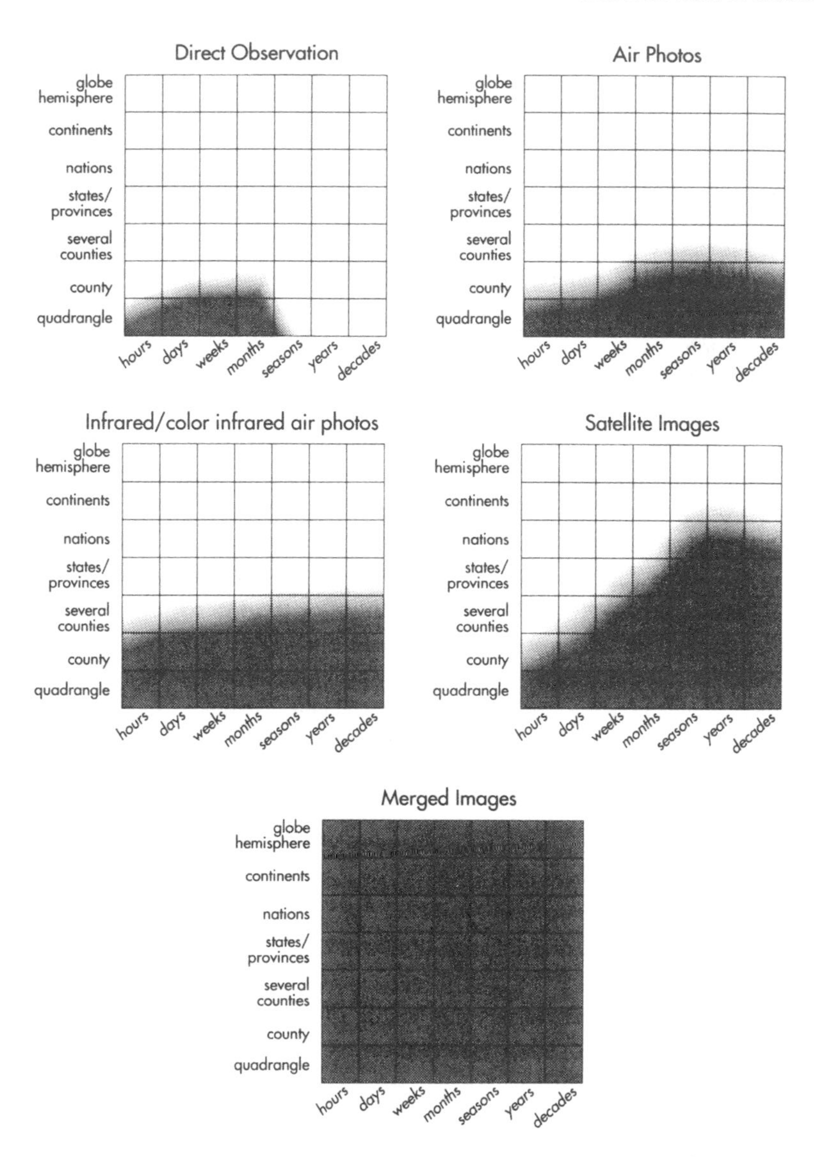

FIGURE 7.1 Geographic data are collected at many different scales by many different methods. This figure illustrates the scale dependence of several data collection methods, ranging from direct observation through fieldwork to merged images obtained from multiple sensor systems on satellites. On each of the plots above, the horizontal axis represents the temporal scale and the vertical axis represents the spatial scale of observation. The shaded areas indicate the range of spatial and temporal scales over which data can be collected for each method of observation. Source: MacEachren et al. (1992, p. 123).

Comparative Studies Using Longitudinal Data

Research on social and biophysical issues must confront two realities: (1) as noted above, processes operating at different spatial scales are intimately but not straightforwardly connected and (2) their outcomes vary widely across space depending on the context (i.e., local conditions) in which they operate. Connectedness and context cannot be addressed adequately unless they are studied together. Moreover, they cannot be understood without considering changes over time, including nonlinear, lag, and feedback effects.

Comparative studies across time are required to advance our understanding of a wide range of geographic problems such as ethnic conflict, demographic change, economic restructuring, and human response to global climate change (e.g., Parry et al., 1988; Mikesell and Murphy, 1991; NRC, 1992a; USGCRP, 1994). Comparative longitudinal studies do far more than demonstrate the existence of variation across time; they are essential to understanding local and regional sources and the implications of globally aggregate (e.g., population growth) or globally systemic (e.g., climate change) processes.

Geography has long engaged in comparative studies through its analyses of place and its analyses of topical or thematic issues across different places (see Chapter 3). Both approaches have been valuable. However, there is a clear need for more systematic comparative case-study analyses, utilizing both quantitative and qualitative approaches, that extend to larger temporal and spatial domains.

Comparative studies of global-scale processes will require sustained and coordinated attention from multidisciplinary teams of geographers and experts from allied natural and social science fields. With a few notable exceptions, geography does not have an established tradition of team research, partly because much geographic research is carried out at more restricted spatial and temporal scales and partly because of a general lack of research support for sustained, team-based work. Adjustments in the research culture and research support systems will be required in order to effectively tackle such work.

To enable comparative studies, methods of geographic representation need to be developed that explicitly address the temporal, as well as the spatial, dimensions of research problems. Research on spatiotemporal representation in the context of GISs and geographic visualization is presently at an early stage (see Chapter 4). Advances in representational theory are needed to incorporate change as a fundamental component of geographic information, rather than as an irregularity that must be smoothed over in order to complete an analysis—as is currently the case, for example, in the enumeration of boundary changes in most GISs. There is a critical need to develop data structures for GISs to support temporal modeling and visualization for longitudinal analysis of global-scale problems. Success in this endeavor will require close collaborations between specialists in representational theory and topical specialists with knowledge of particular problems and information sources.

Impacts of New Geographic Technologies on Decision Making

As outlined in Chapter 4, the past decade has seen a dramatic improvement in technologies designed to store, manipulate, analyze, and display geographic information. The switch from paper maps and tables to digital georeferenced information, including orthophotos and other types of images, has profound implications for decision making.

Geographers have long been active at the "front end" of technological developments, and, in large part because of their efforts, the usefulness and sophistication of technologies have increased significantly over the past decade. Indeed, the sophistication of these technologies has grown to the point that geographic training is necessary to use them to their full advantage. As technology moves from the research laboratory to the workplace, geographers will need to become more involved in the "back-end" of technological applications. The issues here include helping users understand both the potential and limits of technology, helping users understand the impacts of these technologies on decision making itself, and understanding user conceptions of problems to which the technologies are being addressed, as well as of the world being modeled by these technologies. In relation to the latter, geographers need to identify mismatches between the ways in which people understand the world and the ways in which aspects of the world are modeled and presented through geographic technology. More generally, formal models of common-sense geographic concepts are needed as a basis for the design of more intuitive user interfaces to geographic technologies.

Geographers also need to work with users to be certain that new geographic technologies are matched to applications in appropriate ways. Geographers can help users address such practical questions as the following: What geographic models (e.g., location-allocation models) and tools need to be incorporated into GIS software to support decision making? What are the implications for decision making of different degrees of reliability in the data layers of a GIS? How do the visual depictions from a GIS shape decision-making strategies? If geographers do not become involved in these very practical but important issues, there is a risk that appropriate tools will not be used where they can be of help and that inappropriate uses will be made of the tools that are available. It is a high priority, therefore, not only to increase the level of effort being devoted to these back-end issues but also to increase the knowledge base from which this assistance proceeds.

Geographic Learning

The inclusion of geography as a core subject in the Educate America Act (see Chapter 1) reflects a widespread acceptance among the people of the United States that being literate in geography is essential for being an informed and responsible citizen. With the act's emphasis on curricular issues, however, it is easy to forget that little is known about how individuals acquire geographic understanding or which pedagogic approaches offer the most promise (Downs, 1994). Geographers need to become involved in addressing a variety of such fundamental questions about geographic learning. For example, they can help define what it means to be geographically literate, how geography can be taught most effectively over a range of educational levels, and how the effectiveness of learning can be assessed.

For studies of geographic learning to have impact, they must go beyond the single-case examples that currently dominate the literature in geography and education. Geography needs empirical data to address questions about education standards, curriculum design, materials development, teaching strategies, and assessment procedures. More broadly, the discipline needs (1) baseline studies of the current state of geographic education, (2) an agenda to shape a systematic program of research in geographic learning and geography education, and (3) a support system to ensure that this program is carried out and that the results are disseminated.

PROMOTING GEOGRAPHIC COMPETENCY IN THE GENERAL POPULATION

Geography's potential contributions to intellectual and social discourse in the United States cannot be realized solely through academic research. Consideration also must be given to ways of expanding the geographic competency of the general population—many of whom have little idea of what geography is, much less an understanding of its key concepts and tools—and of ensuring that the geographic content and skills being taught are sound. Increasing the competency of the general population will require an expansion of opportunities for geographic learning in elementary and secondary schools, in community and technical colleges, and in four-year colleges and universities and among people who are outside the mainstream educational structure.

The primary role of geography as a science in expanding educational opportunities is to ensure that geographic education and learning are built on a solid base of knowledge. The challenges in this regard are twofold. First, geography as a research discipline needs to communicate its knowledge to teachers and other users as effectively as possible and, when useful new knowledge is developed, as quickly as possible through alliances of professional geographers and teachers. Many such alliances exist now,[7] but they need to be expanded and strengthened. Second, geography as a research discipline must be responsive to users' needs for new scientific knowledge. Research agendas must reflect social needs as

[7] A notable example is the National Geographic Society Alliance Network, which was established in 1986 and now involves some 160,000 participants in all 50 states, the District of Columbia, Puerto Rico, and Ontario (Geography Education Standards Project, 1994).

well as investigator curiosity if geography is to promote widespread geographic competency and associated human well-being.

Geographic Competency Among Primary and Secondary School Students

Nothing is more vital to strengthening the foundations of geography than the improvement of geography education in primary and secondary schools. It has become increasingly apparent in recent years that geography education at these levels is usually woefully inadequate, where it exists at all. An effective response will require that substantially more geographic information and reasoning be taught in and out of the classroom (Geography Education Standards Project, 1994). School-based activities, trips, and clubs can encourage greater knowledge. Outside school, news media, broadcast and print entertainment, self-instruction courses, and computer software can serve the same end. The development of an advanced placement college-entry course and examination in geography could also increase the quantity and level of demand for high school geography education. The greatest challenge is the human dimension: training teachers, many of whom have had no coursework in geography. An additional challenge is incorporating the best geographic knowledge in readily available, understandable educational materials, especially where recent developments are concerned—for instance, on global change issues, most of which have become prominent since the late 1980s.

In devising ways of infusing geography into the curriculum of the nation's schools, it is important to consider ways to reach as broad a spectrum of students as possible. The National Geography Bee draws 6 million participants. The state winners are mostly white males from suburban and rural areas, which indicates the need to make the study of geography more exciting and interesting to females and minority students. The National Science Foundation's support of elementary hands-on science programs may change these patterns, and university-sponsored GIS institutes for aspiring young geographers, both males and females, may help broaden the appeal of geography. Indeed, a minority recruitment initiative launched by the Association of American Geographers in the early 1990s has been highly successful. With support from the U.S. Department of Education, undergraduate minority students participated in summer geography institutes and visited graduate geography programs. More than 50 percent of these students went on to pursue degrees in geography. It is essential to build on these efforts if geography is to break out of a pattern that has led to a substantial underrepresentation of women and minorities in the discipline (Shrestha and Davis, 1989; Lee, 1990; Janelle, 1992).

The rapidly expanding demand for geography instruction in the nation's schools has significant and far-reaching implications for the discipline. Academic geographers will be expected to provide in-service training for current teachers and assistance with curriculum development for school systems, undergraduate and graduate training for aspiring teachers, new and redesigned undergraduate and graduate geography courses to accommodate incoming students with substantial high school backgrounds in geography, and a new generation of teaching and learning materials to support this enterprise.

Geographic Competency Among Community and Technical College Students

As Americans seek to adjust to changing economic and technological circumstances, many are looking to community colleges and postsecondary technical schools for education and training. Many others, daunted by the costs of four-year colleges or seeking to advance their education after years away from school, are flocking to community colleges. Community colleges have often been seen as transitional institutions rather than as sites for basic competence attainment, and technical schools have often lagged behind rapidly changing needs for new training and skills. As a result, many disciplines, including geography, have had limited penetration into these institutions. With so many students looking to community colleges and technical schools for retraining and education, however, and with many institutions broadening their programs to meet more comprehensive educational needs, community and technical colleges could play a significant role in enhancing the geographic competency of the general population.

It follows, then, that an important initiative for geography is to foster more and better geographic instruction in the nation's community colleges and technical schools. At the community college level, programs could be instituted that will help students develop the distinctive geographic competencies needed to function in a global economy and a rapidly changing environment. Technical schools have the opportunity to develop geographic understanding and to prepare students to make effective use of new geographic technologies, including GISs, global positioning systems (GPSs), and computer mapping programs. The goal of these efforts should be to produce a group of community college and technical school graduates who are in a position to think geographically in their daily lives and to employ geographic technologies in specific vocational endeavors.

Geographic Competency Among College and University Students

Both as citizens and as participants in the labor force, university and college graduates are confronting issues and problems that require geographic knowledge and perspectives—ranging from local impacts of global economic change to the effects of changing national demographics on the U.S. economy and the environment. Yet there is little evidence that many of this nation's best-educated citizens are aware of how geographic perspectives would be useful to them.

To address this problem, efforts need to be made to ensure that college and university students have access to geography courses and perspectives that go beyond a concern with "where things are" to provide a basic conceptual and analytical grounding in the spatial and environmental dimensions of human and physical processes. The problems of meeting this demand are formidable, given the current institutional status of the discipline and fiscal constraints that confront higher education in the United States. Despite such institutional obstacles, efforts must be made to increase the availability of geography in the nation's colleges and universities. Otherwise, many of this nation's best-educated citizens will lack the factual background and analytical tools to confront the challenges presented by "a warmer, more crowded, more connected but more diverse world" (Kates, 1994a, p. 1-2).

Geographic Competency Among Those Outside the Education Establishment

A large portion of the U.S. population will be untouched by curricular reforms in the nation's schools and universities: those who have completed their education or who otherwise fall outside the mainstream educational structure. Exposure to geography will help these individuals to exercise their responsibilities as citizens in a nation struggling with problems that have strong geographic underpinnings, ranging from urban crime to the role of the United States in distant conflicts. Training in geography will also promote an appreciation for the physical and social environments in which people live and work and an awareness of the environmental consequences of individual activities and actions.

If this nation is to avoid a lag of at least one generation in public awareness of geography, the discipline of geography must provide opportunities for learning in nontraditional settings such as adult education programs and community centers. The discipline also needs to explore ways of bringing geographic perspectives into government agencies, planning boards, and private businesses, so that decision makers will be aware of the geographic dimensions of the issues they face.

IMPROVING THE TRAINING OF GEOGRAPHERS IN COLLEGES AND UNIVERSITIES

This report has outlined a number of important contributions by geography to scientific and decision-making issues. When these issues are considered alongside major social trends that are shaping the postsecondary educational environment—decreasing faculty-student ratios, the growing diversity of student populations, and the trend toward early specialization—the need for new approaches to training geographers at the college level becomes apparent. The following section focuses on ways to increase the quality of geography training by improving interactions within the discipline and increasing outreach to other disciplines. The last section focuses on teaching approaches and emphases to prepare geographers to contribute more effectively to scientific research and policy making.

Improving Interactions and Outreach

Interaction Among Subspecialties Within Departments

Increased specialization is a well-established trend in science, higher education, and society at large. Such specialization is frequently necessary to prepare students for further graduate and postgraduate training, but it should not undermine one of the most important characteristics of undergraduate education in this country: preparation for a lifetime of critical analysis, flexible work, and continuous learning—characteristics needed urgently in times of rapid change.

Despite geography's fundamentally integrative character, increased specialization has occurred at the expense of the common core learning that once existed across the field's subspecialties in both graduate and undergraduate education. As a consequence, the discipline's strengths in the integration of natural and social sciences in place, space, and time have been lessened, and the discipline's distinctive contributions to liberal education and geographic competency have been concomitantly reduced.

Specialization at the undergraduate level in geography needs to be balanced with opportunities for exposure to different geographic subspecialties in ways that simultaneously reinforce the best qualities of liberal education, prepare students for advanced training, and give all students exposure to geographic research and exploration. The establishment of a more equitable balance between specialization and generalization will require lowering the walls that now exist between subspecialties in many departments.

To this end, several approaches deserve consideration. For instance, a number of geographers have designed courses that cut across the boundaries of traditional subspecialties (e.g., see Sidebar 7.1). These efforts could be built on and disseminated through the geography community. Similarly, physical and human geographers could be encouraged to team-teach courses and to organize seminars and symposia that bridge their subspecialties. Graduate students could be encouraged (or required) to obtain input from faculty outside their subspecialties when designing their thesis and dissertation research programs, and students and faculty could be encouraged to present the results of their research in department-wide colloquia. All of these changes could be made relatively quickly by departments without significant new resources.

Interaction Among Specialists Working in Different Universities

Faculty size in the top departments of geography in the United States is small relative to other natural and social science disciplines (see footnote 3),

SIDEBAR 7.1 Hypothetical and Real Worlds

To teach, I would build a trap such that, to escape, my students must learn.

—Robert M. Chute, *Environmental Insight*, 1971

Since the early 1980s, an introductory geography course at the University of Minnesota has placed students in a trap, triggered by a very simple device: on the first day of class, students must relocate the poles (i.e., the axis of rotation) of the Earth (see Figure 7.2). They thereby change a host of factors relating to the biophysical patterns of the Earth and the sorts of human settlement and livelihoods that people might fashion from it. Yet the possibilities for human endeavor lie not only in those provided by nature but also in the way in which people conceive what is possible.

The purpose of this course is to introduce undergraduate students to the ideas and models that geographers use to study the character and relationships between biophysical and social processes on the face of the Earth. The course emphasizes the ideas and presuppositions underlying "facts" about the world, not the facts themselves. It highlights unstated assumptions, shows the importance of empirical information for analysis and decision making, and increases student consciousness of scale and the intimate relations between scale and modes of explanation. Through discovery, discussion, and individual and class exploration, students are able to reason out probable and possible patterns of the hypothetical world and thereby understand something about the real world, as well as about the models and ideas geographers use to study and explain the real world.

Students work cooperatively in groups of six to eight to produce an atlas of the hypothetical world consisting of some 24 maps. The course ends with a "debate" in which each group presents its hypothetical atlas for discussion. Students find that the hypothetical world assumes an amazing reality for them; they understand how and why geographers ask the questions they do, and a number decide to major in geography and even to choose it as a career.

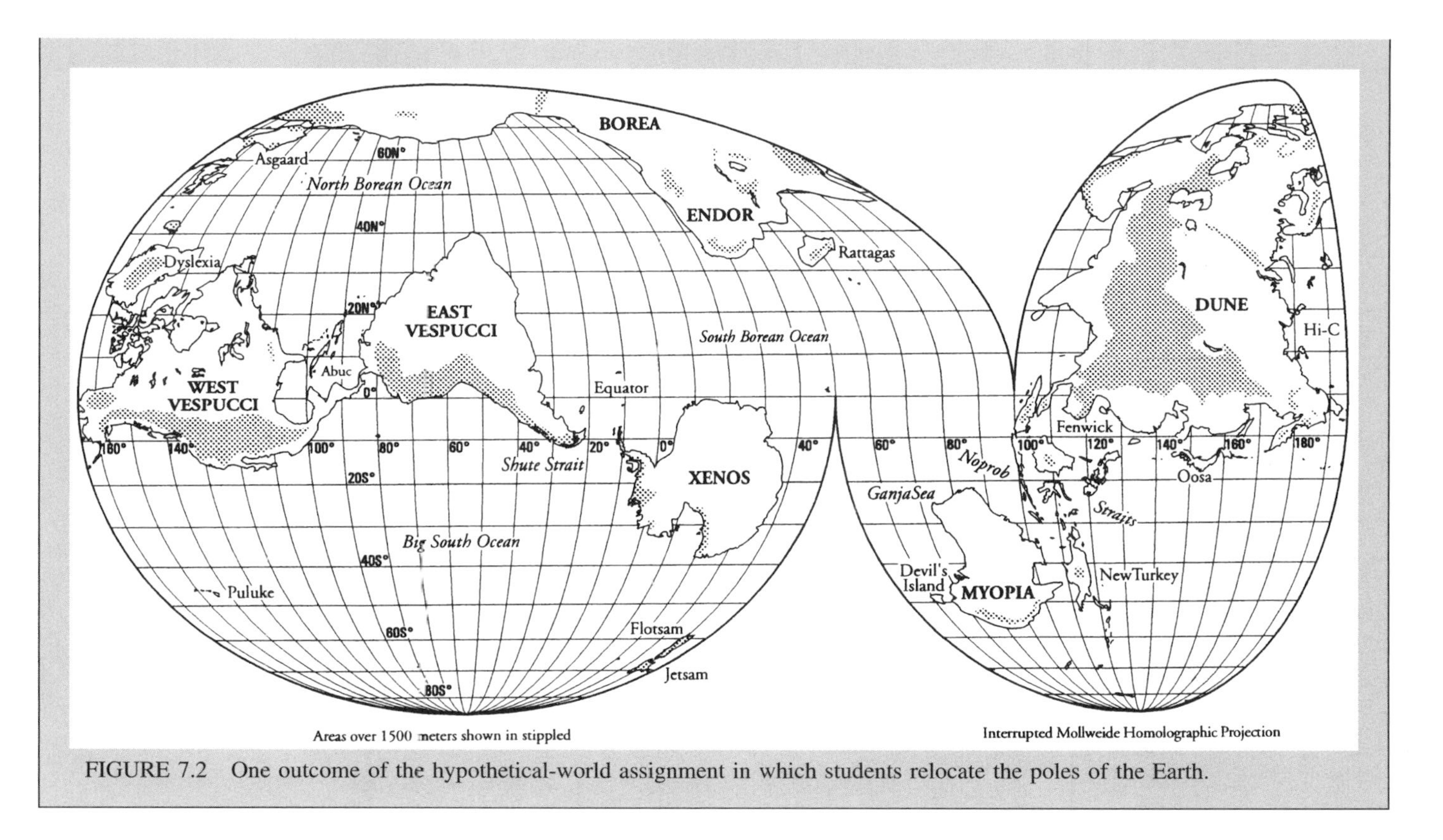

FIGURE 7.2 One outcome of the hypothetical-world assignment in which students relocate the poles of the Earth.

despite the fact that the breadth of subject matter of geography is as large as, if not larger than, that of most other disciplines. As a consequence, geography departments have often had to sacrifice depth of coverage in selected subdisciplines in order to maintain breadth of coverage across subspecialties. Yet depth of coverage is essential to conducting cutting-edge research, particularly in departments with Ph.D. programs. One way to obtain depth without increasing department size is to emphasize more team-based, interdisciplinary work within and across institutions, which many geographers now accomplish through interdisciplinary institutes and centers. Another way is to adopt procedures used by organizations such as the National Center for Geographic Information and Analysis, in which groups of fellow specialists from different institutions are established to overview and guide specific research ventures.

Diversity in Students and Perspectives

Colleges and universities throughout the United States are witnessing several dramatic changes in their students: they are older, they often transfer from community colleges, and they have more diverse backgrounds. Very shortly, these individuals also will bring with them far better precollege preparation in geography, and they will be more proficient in the use of information technologies. To provide role models for this increasingly diverse group of students, colleges and universities will be hard pressed—for example, to tap an extremely limited pipeline of talent among minorities. This underrepresentation of minorities in the discipline (see footnote 6) is mirrored in the present student population. Yet in the years ahead a growing number of minority students will be coming from schools with enriched geography offerings, as a result of several developments: the new emphasis on geography as a core subject in precollege education (see Chapter 1), the implementation of precollege geography standards (Geography Education Standards Project, 1994), and more effective teacher training provided by alliances such as the National Geographic Society Alliance. These students will bring with them a mind-set attuned to concerns about the environment; they will be comfortable with CD-ROMs; they will be familiar with cyberspace; and, perhaps most importantly, they will be increasingly sensitive to issues of diversity, whether they come from New Delhi, downtown Baltimore, Nairobi, Quito, or Miami.

Geography has a tremendous opportunity to benefit from the diverse perspectives and backgrounds that these students will bring to higher education. Yet much of geography is not yet ready to embrace this diversity, despite the substantial headway that critical social theory has made in academic human geography. Too often, the geography that is taught in colleges and universities places "otherness" in contrast to middle-class, temperate-zone, North American norms. This tendency is by no means unique to geography, but it has the potential to alienate some students from formal or informal study of the field.

If geography is to draw on the breadth, talent, and insights of students attending U.S. colleges and universities today, it must devise ways of reaching out to those from "other" communities. The geographic approaches outlined in this report have the potential to engage a broad range of individuals and give them the power to interpret and understand environmental, economic, and political events that affect all of humankind. If the growing number of these well-prepared students can be recruited for geography courses, they could bring new and important insights to studies of problems that have a profoundly significant impact on all peoples.

Geography's concern with diversity must extend to women as well. The number of women attracted to the discipline has risen markedly in recent years, but women are still greatly underrepresented in more senior faculty positions (footnote 6). There must be a continued commitment on the part of faculty and administrators to recruit and promote qualified women to senior positions.

Improving Teaching and Learning

Interactive Learning Technology

College students currently taking introductory geography share one commonality: many have never had a geography course and know little about the field. Most precollege teachers lack strong preparation in geography. If they are to become more effective geography teachers, they will need new teaching tools to help them incorporate geography into their courses. Computer and telecommunications technologies provide powerful interactive learning tools to address this need. Such tools can help teachers develop the background knowledge necessary for effective instruction and can give students a chance to learn and experiment at their own pace and in creative ways.

Efforts are already under way to develop interactive learning tools for instruction at the high school level (see Sidebar 7.2). Such tools have the potential to help fill a major gap in the U.S. education system—the paucity of trained geography teachers—by offering approaches to learning that embed instruction in computer software. Although interactive learning has the potential to enhance and individualize learning in all disciplines, it has particular applicability to instruction in geography owing to the spatial and scalar nature of many geographic problems (see Chapter 3). The traditional concern with space and place requires effective cartographic depiction along with other visual representations such as graphs, photos, and remotely sensed images. Geography also emphasizes processes occurring in and across space and through time. Process is particularly hard to illustrate with words and static images. The technology that makes interactive learning modules possible also facilitates the dynamic depiction of process.

Interactive learning tools are most effective when they are coupled with multimedia classrooms that allow teachers and students to work as a group.

SIDEBAR 7.2 Interactive Tools for Geographic Instruction

During the past few years, several computer-based interactive learning tools have been developed for geography instruction. This sidebar illustrates two particularly noteworthy examples. The Encyclopedia Britannica Educational Corporation is producing the *Britannica Global Geography System* (BGGS), a series of computer-based instruction modules for each major world region that were developed by the Center for Geographic Education at the University of Colorado at Boulder. The BGGS is comprised of 20 modules, each of which focuses on a single geographic issue. Examples include hunger, pollution, natural hazards, political change, development, regional integration, and nation building. Modules were developed by combining the pedagogical expertise of veteran geography teachers with the content expertise of geography scholars. Each module offers a case study in one of 10 world regions and also shows how the issue plays out in North America. The modules also provide suggestions to teachers on how to bring the issue to the local level. For example, the module on hunger suggests that a local soup kitchen operator be invited to class to talk about hunger.

Another interactive learning tool is being developed by a multidisciplinary team of geographers, computer scientists, teacher educators, and learning resource specialists at Virginia Polytechnic Institute and State University. This tool, GeoSim, consists of a series of five computer modules: Human Population, Migration and Political Power in the United States, Migration and Sense of Place, Migration Modeling of the United States, and Maps and Mapping. Each module consists of a multimedia tutorial and a simulation exercise (which deal with topics such as international migration or political power) that can be used together or separately in class, as homework, or as a laboratory exercise. The modules, which combine capabilities of GIS and computer simulation, emphasize interactive learning.

The modules allow students to manipulate parameters of the simulations and then observe the simulation play out through a variety of maps and diagrams (see Figure 7.3). The modules emphasize repeated trials that allow students to test and refine hypotheses. Project GeoSim software is available via anonymous FTP or Gopher at geosim.cs.vt.edu. It can also be accessed via the World Wide Web at http://geosim.cs.vt.edu.

Traditional presentations of technical topics such as map projections can be infused with life by the simple expedient of having the spherical Earth transform into flat maps of various projections before students' eyes. Explanations of the Coriolis force in physical geography or processes of disease diffusion in human geography can be made immediately comprehensible through dynamic visual presentations. The same multimedia classroom tools can be used to instruct students in the use of interactive learning modules that build on classroom presentations.

Regional Knowledge and Expertise

What are the major social ramifications of economic restructuring in Eastern Europe? What is the future of democratization in Latin America? What forces

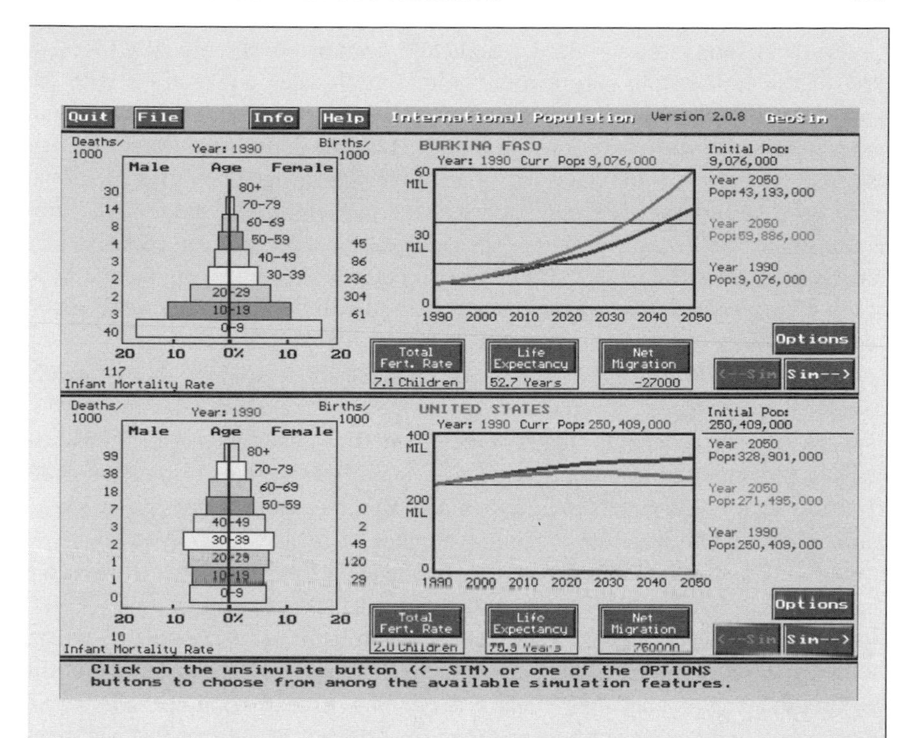

FIGURE 7.3 Sample screen from GeoSim's International Population Module. GeoSim is produced by the Departments of Geography and Computer Science, Virginia Polytechnic Institute and State University, Blacksburg.

are encouraging the growth of Islamic fundamentalism in the Middle East and North Africa? Questions of this sort are of concern to both scientists and decision makers, and broad-based efforts to address them often draw on the expertise of regional specialists.

The long-standing tradition of foreign-area research and foreign-area fieldwork in geography has served the discipline well in terms of its involvement in regional study programs at colleges and universities. It is less clear whether this tradition has been as successful beyond these institutions. Geographers appear to be less well represented on major interdisciplinary boards and committees on geographic regions, such as the Social Science Research Council (SSRC)[8] or in

---

[8]The SSRC organizes workshops and symposia and funds joint research projects of scholars working on regional problems.

government-sponsored activities specifically designated as regionally focused. Part of the explanation may involve geography's weak connections with the political science community, which strongly influences certain kinds of regional programs, and with institutions of higher learning that have historically fed the U.S. Department of State and other government agencies using regional expertise. At any rate, few geographers have participated in the regional programs of the SSRC, and individuals with geographic training are not consistently represented on SSRC panels. They are also conspicuously lacking from a wide range of other federal and private scientific and policy initiatives that focus on regional issues.

Another reason for geography's poor representation may be that relatively few early-career U.S. geographers define their specialties in regional terms. Foreign students account for many of the dissertations that focus on foreign areas in U.S. geography graduate programs. Substantial percentages of these students return to their home countries after completing their studies (Turner and Varlyguin, 1995). Moreover, the declining number of articles submitted to major professional journals that deal with developments in foreign regions (Rundstrom and Kenzer, 1989; Brunn, 1995) suggests that students doing foreign-area dissertation research switch to other topics after graduation or focus on topical problems at the expense of larger regional developments. Whatever the causes, one of the important learning challenges facing geography is to expand the number of geographers who can contribute to broad-based scholarly and policy initiatives focused on regional issues.

In pursuit of this end, geography needs to consider ways of nurturing students' interest in regional developments and problems and of honing their abilities to analyze such problems using geographic concepts and tools. There are two intertwined needs here. One is to train a significant contingent of geographers in the history, languages, institutions, social practices, and environmental processes of particular world regions so that they can address a wide spectrum of regional issues and problems. In many cases, of course, this raises a further need: financial support in order to spend periods of residence in a region. The second need is to bring the concepts and analytical insights of geography to bear on such regionally focused, multidisciplinary issues as ethnic strife, global economic change, and urban population growth in developing countries.

Geographers are already beginning to respond to these challenges. Interest in regional geography is on the rise, as evidenced by the growth in membership of some of the regionally focused specialty groups of the Association of American Geographers. If these trends continue, geographers are likely to play a greater role in the years ahead in regional programs such as those of the SSRC. It is far from certain that this will happen, however. Geographic considerations are likely to remain marginalized in many efforts to comprehend regional issues unless there are a substantial number of geographers with sufficient interest and training to understand particular regions—and to raise evocative geographic questions about them. In turn, such efforts would be impoverished by the lack of geographic insight.

Field Exploration and Discovery

The recent development of global databases and remote sensing technologies has tended to direct attention away from fieldwork as an element of geographic research. Rather than making fieldwork obsolete, however, these developments have led to new questions about the meaning and significance of data, about the geographic patterns that underlie and influence data, and about the gaps and shortcomings in databases and technologies for data acquisition (see Chapter 4).

Fieldwork is not only important as a check on datasets and remote sensing imagery; there are enormously important forces at work shaping the geography of the planet that are unlikely to be understood unless one ventures into the field. Many critical issues in physical geography cannot be addressed by using existing data sources and imagery. For instance, most of what is known about vegetation history during the Quaternary, the dynamics of forests, and the behavior of stream channels comes from systematic fieldwork. Moreover, additional field studies offer the greatest potential for expanding our understanding of these topics in the future. In the human arena, for example, no data sources exist on the role of the "overseas Chinese" in the economy of Southeast Asia or the cultural norms that lead various groups in the Nepalese highlands to adopt different land-use strategies. Sole reliance on secondary data sources for investigating social and political developments can produce significant gaps and omissions in research and understanding. Because most economic and political data are nation specific, the development of links between regions on either side of international boundaries—for example, in the Mekong River basin—is likely to be overlooked if geographers rely solely on published data. Most fundamentally, of course, the kinds of spatial variations addressed by geographers are more complex than the boundaries that define most available data.

There is a real need, then, for geography to build on and expand its tradition of fieldwork. Students need to be exposed to field studies and encouraged to develop their powers of observation and analysis in the field. Beyond its importance for research, encouraging fieldwork has several potential side benefits. One is to engage student interest. Many academic geographers were attracted to the discipline by its real-world focus, its concern with what one geographer has called "the world you can kick." Incorporating field studies into geography courses is likely to attract students who are interested in connecting the world of the mind with the concrete world in which they live. Moreover, opening up field experiences to students can enhance their understanding of the complexity of the world and heighten their appreciation for the geographic consequences of decisions and actions.

New Geographic Technologies and Data Sources

The discipline of geography is in a period of transition from a past where geographic information was transmitted in the form of paper maps to a future in which most geographic information will be transmitted through digital information systems. The early signs of this transition can be seen in the dissemination of satellite images of cloud cover and other meteorological/climatological data on the Internet. A stream of new technologies such as GPSs will provide georeferenced information on a great variety of activities about which little is presently known. Vast quantities of new digital information about the Earth and its peoples are becoming available, and national and international standards for spatial data exchange are being developed and adopted. Geographers need to learn about these spatial information sources and understand how to use them appropriately in their work. Academic geographers also need to consider how new technologies and data sources will create new opportunities for collaborations with the private sector.

If geography is to meet the challenges of rapid technological change, steps must be taken to familiarize students with new technologies for data analysis and display. The past two decades have seen an explosion in computing power and the ability to process and store data. Technology now plays a preeminent role in a wide range of geographic research. In the environmental arena, for example, parallel processing technologies have opened new possibilities for addressing computationally intensive problems such as global climate change. A broad array of data representing the physical world consists of point samples representing surface or volumetric phenomena. Geographers, and others, have developed successful algorithms for interpolation of surfaces and volumes from such data. Continued advances of this sort will require a substantial coterie of geographers who understand new technologies and can use them effectively.

There are also general computer hardware, software, and networking issues that point to the need for increased technical competence in the discipline. Geographers have come to rely on digital communications networks to support both research and instructional efforts. They also actively use this technology to disseminate information to the broader science and applications communities. For example, the geographic information system listserve (GIS-L) is reported to have a national and international readership exceeding 20,000—almost three times the size of the Association of American Geographers. As the use of digital libraries and such digital information exchange tools as the World Wide Web becomes more common, the importance of technological literacy in the profession becomes obvious. Geographers need to be trained to understand the logic of data processing and networking so that new technologies can be exploited to their maximum potential in the service of geographic research and teaching.

8

Rediscovering Geography: Conclusions and Recommendations

The preceding chapters have offered a number of observations about, challenges to, and strategies for realizing geography's potential to contribute to scientific understanding and societal problem solving in the United States. In this final chapter the committee summarizes its conclusions and recommendations on steps to improve geographic understanding, improve geographic literacy, strengthen geographic institutions, and take individual and collective responsibility for strengthening the discipline.

IMPROVING GEOGRAPHIC UNDERSTANDING

Clearly, geography has too few answers to the questions being posed to it by society, although its potential to answer those questions is considerable. At the same time, geography is being asked too few questions by the other sciences. On the one hand, the demands of *society* are too large for the current capabilities of the discipline; on the other hand, the demands from *other scientific disciplines* are too small. Because geography's ability to respond to society's needs depends considerably on its strength as a science, and its strength as a science depends considerably on its support from the family of sciences, this contradiction is a matter of serious concern to the committee.

Given society's enhanced interest in geography as a subject, it is essential to improve the knowledge base of geography as a discipline related to critical issues for science and society, to increase the appreciation and use of geographic perspectives in science and society at large, and to treat geographic learning as a challenge to science as well as to practice. Based on the foundation laid in Chapters 1 through 7, the committee concludes that responses by the discipline and by its external constituencies are needed to:

- *Improve geographic analysis in a new era of data and analytical tool availability, related to broader needs of science.* Geography has made remarkable advances in its analytical capabilities during the last generation, but it faces significant challenges in responding to the emergence of new data types and analytical needs. For instance, certain types of georeferenced data (e.g., census tract data) are now available in such great and rapidly expanding quantities that they threaten to swamp thoughtful analysis, especially by untrained users. Moreover, the availability of large quantities of data tends to mask a broader underlying problem: namely, that data availability is not always well matched to data needs. Improved capabilities for data collection and analysis should therefore be high on the discipline's research agenda.

Geographers must also improve the practice of relating the ''front end'' of geographic analysis—conceptualization and data selection/sampling design—with ''back-end'' modeling and analysis. Without thoughtful and intellectually robust linkages between these two elements of the research enterprise, geographic analysis will be inherently incomplete. At least as important, the capacity of geographic analysis to address issues of complex systems and nonlinear dynamics needs to be improved in order to fulfill geography's potential to contribute to the body of science. The improvement of capabilities for data collection and analysis should therefore be high on the discipline's research agenda.

In addition, it is important to recognize the value of utilizing a variety of methodologies in seeking better understandings of the world, combining geography's characteristic appreciation of diversity with its recognition that there is no single ''foolproof method'' for producing knowledge.

A particular challenge is that of analyzing and modeling relationships among natural science and human science phenomena and processes, which are so often separated by boundaries of epistemology, professional specialization, data categories, and units of measurement. Besides the technical challenges, such as relating economic and ecological indicators, this is also a challenge to individual scientists to transcend conventional boundaries for understandings of other kinds of processes and linkages.

- *Develop integrative, interdisciplinary, relatively large geographic research initiatives in response to priorities of science and society.* If geography is to increase its contributions to science and society, it must learn to think more broadly and to respond to science agendas set beyond the confines of the discipline. When this course has been followed, the utilization of geography's perspectives and knowledge base has increased immeasurably. The National Center for Geographic Information and Analysis, for example, has reinforced geography's core role in a mode of analysis and techniques that are not only at the forefront of various national and international agency agendas (e.g., National Aeronautics and Space Administration, Census Bureau) but that are also being adopted throughout many natural and social sciences.

Geography as a discipline should devote more attention to the development of larger, integrative, interdisciplinary research projects, particularly projects that would benefit from the collaboration of physical and human geographers or those who develop methods of spatial representation and those who apply those methods, both within the discipline and beyond; and more of this research should be directed at high-priority issues for society and science. An example of a research issue suited to both of these emphases is global change, broadly defined (see Chapter 7). In this and other cases, geography's ability to contribute on the basis of sound scientific research will often depend on the availability of valid longitudinal information covering the diverse topics incorporated in the discipline's perspectives.

• *Increase the use of geographic perspectives to provide scientific insights that may not be achieved in other ways.* Geography's spatial approaches (see Chapter 3) increasingly influence research in many fields beyond geography. In addition to geography's contributions to subject-matter knowledge as such, its way of thinking and its skills in understanding visual representation should also be utilized more often throughout science to improve scientific understanding. Individual geographers should therefore become more engaged in interdisciplinary research activities that bring such geographic perspectives and tools to bear on important scientific and societal questions.

• *Increase linkages between geographic research and geographic education, by emphasizing research on geographic learning.* If geography is to be an effective contributor to improving the knowledge base in schools, and reach out to America's adult population, it needs to undertake research on how geographic learning takes place. Besides its general value for education, such research will help assure that geography's perspectives and skills are used by colleagues in other sciences, who often are engaged in geographic learning themselves. Research attention also is needed to address what geographic literacy means and how it can be facilitated. The results of such research will strengthen decision making related to education standards, curriculum and materials design, and assessment.

IMPROVING GEOGRAPHIC LITERACY

Geography is being asked by the nation to help improve the geographic literacy of the U.S. population: knowledge of the world and flows within it, characteristics and dynamics of places, relations between local and global changes, relationships between people and their environment, and uses of geographic data and capabilities for data display and analysis. To respond effectively, the committee concludes that steps are needed to:

• *Implement programs to support and assure the quality of the science content in kindergarten through grade 12 (K–12) geography education.* Although the committee did not evaluate the current state of geography education in America's schools, it shares the widespread impression that this education is not up to world standards (e.g., see Chapter 1). If geography education is to receive substantially more attention in the nation's schools, it is important for geography as a discipline to assure that the educational content is sound and current. This calls for initiatives to improve both teacher training and educational materials, with specific attention to student performance and curriculum.

• *Foster conceptually sound general education courses in geography as part of a liberal arts college education.* Beyond the K–12 level of education, geography should be contributing more effectively to the training of students at the college level, whether or not they are geography majors. Courses in geography should emphasize the discipline's perspectives and skills as well as its subject matter. An American student with a liberal arts higher education, whether from a university or a technical/community college, should be geographically competent—knowledgeable about the world and able to use geography's perspectives and skills in the workplace and in life.

• *Develop programs that bring geographic perspectives to bear more effectively on business, government, and other organizations at national to community levels.* Clearly, the United States cannot wait a generation—until geographically literate students move into positions of responsibility—to utilize geography to improve decision making and social well-being. Improvements are needed now. To this end, improved linkages are required between geography's professional practitioners and prospective users of its perspectives in business, government, and other organizations at all scales.

• *Use existing institutional bases to expand access by the U.S. population to geography.* The effective use of geography's perspectives by decision makers in this democratic society will depend substantially on a higher level of geographic literacy throughout the country's population. This challenge involves at least two elements: getting the information out and getting people to use it. For geography as a research discipline, the main responsibility is getting the science right, in terms of substantive content and interactive communication. But geographic ignorance in the United States probably cannot be reduced significantly without strong collaboration among a variety of concerned parties, including geography's organizations, government at several levels, business, nongovernmental interest groups, and the information media. In particular, progress is likely to depend on whether or not the nation's private sector sees geographic literacy as a business opportunity. Strong linkages are needed between professional geography and the business firms that might undertake such an effort, in order to reduce the costs to business of getting geographic information and insights into the marketplace rapidly and to assure that the products and services are sound as well as attractive.

STRENGTHENING GEOGRAPHIC INSTITUTIONS

To meet its scientific and societal responsibilities, as outlined in this report, geography as a discipline needs to change, with the support and encouragement of its institutional patrons. The most fundamental problem is one of magnitude: geography's small size relative to demands for its services, broadly defined. But the discipline needs to address itself to problems of substance as well, related both to its traditions and to new directions in response to changing conditions. To these ends, the committee concludes that initiatives are needed to:

- *Rediscover traditional strengths of geography.* Much of the recent external interest in geography has been focused on disciplinary traditions that are considered by many geographers to be more characteristic of the past than the future: integrative knowledge, regional knowledge, and field discovery. Geography as a discipline needs to reexamine these traditional strengths, reconsider their intellectual and societal relevance, and expand the attention devoted to them in research and teaching.
- *Discover and pursue new directions for geography.* As geography rediscovers its traditions, it also needs to focus on new directions that are essential if the discipline is to remain relevant into the next generation. Directions needing increased emphasis include connections with critical issues for society, involvement in intellectual challenges to science at large (e.g., analyzing and understanding complex dynamics), and the pursuit of opportunities for interactive learning as a challenge for both research and teaching. Several of these directions were addressed earlier in connection with programs to bring geographic perspectives to bear in business, government, and other institutions.
- *Expand geography's resources and reach, reconciling supply and demand.* All of the conclusions listed above ask that geography do more in the future than in the past. The most profound problem, and in many ways the most urgent, is that geography is being asked to increase its contributions to science and society at a rate unprecedented in the discipline's history in the United States, yet it remains a small academic discipline situated in educational institutions with very limited capacities for expansion. In terms of geography's reach, it is being asked to strengthen K–12 education and improve the geographic literacy of the general population and major decision-making institutions. In addition, its perspectives and skills have new relevance for many multidisciplinary research goals and for science in general, and the discipline needs to match expanded educational roles with expanded research roles in order to maintain a healthy balance as a science. But its human and financial resources in universities, departments, and external sources of research funding are painfully limited—and already strained nearly to the limit in many cases. One key to resolving this mismatch between reach and resources is governmental support, especially as a catalyst for change in the near term (support that should not be equated simply

with more money). In the longer term, however, the key may be to attract other external sources of support, including foundations and the private sector.

Progress in these regards is likely to depend as much on initiatives arising from within the discipline than on external initiatives. Simply stated, geography needs to do a better job of identifying users of its knowledge and techniques—and in estimating demand levels, demand trends, supply priorities, and supply strategies—in order to develop strategies to increase its external resource support. Considering the critical roles of external demands and resources in shaping such a prospect, however, this vision cannot be developed by geography in isolation. As indicated in Chapter 1, geography is a means to social ends, not an end in itself; and its plan for expanding its resource base must be consistent with those ends and the societal resources allocated to reach them.

An example of a specific issue is the growing dependence of effective geographic research and teaching on capital equipment. Except for physical geography and cartography, geography departments in colleges and universities have generally not needed significant equipment budgets in the past. At least partly as a result, the impacts of the technological revolution described in Chapters 4 and 7 have been virtually impossible to accommodate within current institutional concepts of departmental budgeting. In addition, the shorter effective lifetime of higher-technology equipment calls for budgets for regular replacement as well as for base-level capital stock.

• *Alter faculty reward structures in universities, colleges, and geography departments.* Reward structures need to recognize the importance of long-term, collaborative research; geographic education, including scholarly contributions to geographic learning; contributions to societal problem solving; and interdisciplinary interaction, including working and publishing with colleagues in other disciplines.[1]

TAKING INDIVIDUAL AND COLLECTIVE RESPONSIBILITY FOR STRENGTHENING THE DISCIPLINE

Finally, geographers need to recognize that they also have responsibilities to their discipline, to other sciences, and to society. Geography's new relevance does not just pose challenges to external bodies and larger institutional settings; it calls for a response by individuals and groups of geographers on their own volition, both as professionals and citizens. In particular, the committee suggests that geographers, their organizations, and their departments should:

• *Recognize education and service as professional responsibilities.* In line with general trends in the world of higher education, geographers need to respond

[1] See Association of American Geographers. 1994. Reconsidering Faculty Roles and Rewards: Washington, D.C.

to societal needs for quality education and practice as professionally as they respond to opportunities for research, showing sensitivity to societal need as well as self-directed curiosity.

• *Enhance diversity in geography's perspectives, participants, and audiences.* Geographers and their organizations need to make themselves fully aware of trends in the diversity of student and professional populations and trends in the diversity of approaches to seeking understanding. As an expression of the nature of geography, as well as morality and humanity, they must appreciate diversity, value it, support it in their institutional environments, and seek it in their own learning. Furthermore, geographers need to endeavor to address research issues of relevance to a wider range of user communities, including disadvantaged groups.

• *Promote breadth and depth of learning, emphasizing the common core of learning that provides coherence to the discipline.* Geography faculty need to take collective responsibility for identifying and infusing into undergraduate and graduate programs the core of conceptual and methodological approaches that provide coherence to the discipline. They need to ensure that undergraduate and graduate students are solidly grounded in the fundamentals of geographic learning, while at the same time affording graduate students opportunities for in-depth training at the frontiers of knowledge in selected subspecialties.

To this end, faculty members need to make more of an effort to collaborate across subspecialties, especially between human and physical subspecialties, between these subspecialties and those emphasizing spatial representation, and across institutions. Faculty need to ensure that their undergraduate and graduate students are exposed to the range of geographic topics, research traditions, and methodologies that define the core of the discipline. At the undergraduate level, faculty members need to provide exposure to the various subspecialties in a way that reinforces the ideal of a liberal education and at the same time prepares students for advanced training. Such exposure might come, for example, through courses and seminars led by teams of physical and human geographers that emphasize the connectivity among subspecialties. At the same time, faculty members must provide opportunities for graduate students to obtain depth of learning in selected subdisciplines. Such opportunities might come through pre- and postdoctoral training opportunities, collaborations with faculty members in other departments, and extracurricular activities such as summer institutes.

• *Promote and participate in professional interactions with other sciences.* As participants in a larger intellectual enterprise and as individual representatives of geography in an era of reaching out more actively beyond disciplinary boundaries, geographers need to seek interactions with other scientists, not only through multidisciplinary programs and research projects but through a wider range of discourse. This process might, for instance, include more active participation in state academies of science and such national organizations as the American Association for the Advancement of Science, as well as catalyzing discussions of cross-cutting scientific issues with colleagues in one's own institution.

• *Recognize responsibilities to local partners in conducting field research, especially in foreign areas.* Field research in geography, which needs to receive increased emphasis in the years ahead, carries with it a number of obligations. Besides care in the science, such as sampling and inference, geographers need to recognize that local expertise is usually an essential part of thoughtful field research and that the benefits they draw from that enterprise often call for something of value to be returned. For example, research results of local or regional interest might be reported to diverse local audiences for their benefit, and/or contributions might be made to building local capacities for doing such research. As a general rule, U.S. geographers conducting research throughout the world need to work in collaboration with local counterparts.

RECOMMENDATIONS

Based on these conclusions, the committee offers a number of recommendations addressed to the external audience of this report. Each recommendation addresses the question of who needs to take action, and the final recommendation addresses the implementation process itself. Professional geographers should also examine the conclusions on "taking individual and collective responsibility for strengthening the discipline" of geography in order to identify additional actions that could strengthen the discipline from within.

To improve geographic understanding:

1. **Increased research attention should be given to certain core methodological and conceptual issues in geography that are especially relevant to society's concerns.** Key issues include complex systems and nonlinear dynamics, relationships between physical and human geography, multiscale analysis, comparative case study analysis, and visual representation. Implementation responsibility lies with those institutions that fund research in geography and related fields, along with universities, professional societies, and other institutions that promote the generation of research ideas and proposals (as well as geographers themselves).

2. **More emphasis should be placed on priority-driven, cross-cutting projects.** A larger portion of geographic research support should be allocated to multiple-investigator, multidisciplinary projects that address scientific and societal priorities such as global change, urbanization, conflict resolution, and the dynamics of complex systems, with implementation by the same institutions as in the previous recommendation. Geography's organizations should be actively involved as catalysts for project and group development, especially involving more than one institution, and to assure that large research groups are accessible to, and are in communication with, the larger research, educational, and other user communities.

3. **Increased emphasis should be given to research that improves our understanding of geographic literacy, learning, and problem solving and the roles of geographic information in education and decision making, including interactive learning strategies and spatial decision support systems.** This recommendation calls for a new collaboration between research support institutions such as the National Science Foundation (NSF) and parties directly involved in teaching, learning, and other applications of geographical knowledge and tools. Geography's organizations—the Association of American Geographers (AAG), the American Geographical Society, the National Council for Geographic Education, and the National Geographic Society—should take the lead in fostering this collaboration and proposing strategies to enhance it.

To improve geographic literacy:

4. **Geography education standards and other guidelines for improved geography education in schools should be examined to identify subjects where geography's current knowledge base needs strengthening.** Examples where strengthening is needed are likely to include such traditional strengths of geography as the pursuit of scientific synthesis through integration in place, addressing issues of relevance to the diversity of social and economic groups that constitute U.S. society, approaches to relating direct field observation to modern information technologies, and foreign field research. The responsibility for identifying critical gaps lies with geography's organizations.

5. **A significant national program should be established to improve the geographic competence of the U.S. general population, as well as leaders in business, government, and nongovernmental interest groups at all levels.** A major multiyear effort is needed to assure that the knowledge, perspectives, and skills of geography as a subject are utilized effectively in meeting such national needs as competitiveness in the global economy and sustainable democratic responses to issues and choices in government, including foreign policy, environmental policy, and information infrastructure policy. This effort should be established through cooperation between the U.S. federal government, the NGS, and other geography organizations.

6. **Linkages should be strengthened between academic geography and users of its research.** Geography's organizations should increase their interactions with private-sector firms and associations, government agencies, educational institutions, nongovernmental interest groups, and appropriate funding organizations to examine ways to improve the effectiveness of information and technology transfer, to increase personal and professional linkages, and thereby to improve business and government and community decision making.

To strengthen geographic institutions:

7. **A high priority should be placed on increasing professional interactions between geographers and colleagues in other sciences.** Geographers

themselves bear a large part of the responsibility for this collaboration; but research support institutions and other institutions, such as NSF, the National Research Council (NRC), and the Social Science Research Council, which appoint groups to address issues where geographic perspectives and expertise are relevant, should also contribute to the implementation of this recommendation. The AAG is well suited to lead this effort.

8. **A specific effort should be made to identify and address disparities between growing demands on geography as a subject and the current capabilities of geography to respond as a scientific discipline.** Specific recommendations concerning funding increases go well beyond the knowledge of committee members about scientific issues, but this issue urgently needs attention. Moreover, it is an issue that cannot be adequately addressed without the participation of a wide variety of institutions and constituencies, including government agencies, private foundations, school boards, institutions of higher learning with existing geography programs, and institutions that lack those programs. The committee recommends that geography's organizations, universities, the NRC, the NSF, and other appropriate agencies join together to produce a careful assessment of the resource constraint issue as well as the implications of different scenarios for resolving it.

9. **A specific effort should be made to identify and examine needs and opportunities for professional geography to focus its research and teaching on specific problems or niches, given limitations on the human and financial resources of the discipline.** As a parallel effort to the previous recommendation, geography's organizations—particularly the AAG and the NGS—in collaboration with research support institutions such as NSF, should consider ways of using their scarce resources for maximum scientific and societal benefit by prioritizing, focusing on especially important needs and opportunities. This prioritization need not, and should not, be at the expense of individual investigator curiosity, but it should stimulate that curiosity by offering a vision of scientific accomplishment and impact related to especially promising directions for contributions by professional geographers.

10. **University and college administrators should alter reward structures for academic geographers to encourage, recognize, and reinforce certain categories of professional activity that are sometimes underrated.** Such categories include long-term research; collaborative research; research directed toward societal problem solving, including policy research; research on geographic learning; geographic education as a field of scholarship, teaching, and professional service; and interdisciplinary interaction and communication. This recommendation should be implemented by university and college administrators, in collaboration with geography's organizations and with national and state associations concerned with reward systems and personnel policies for university faculty members.

To encourage implementation of these recommendations:

11. **Geographic and related organizations—especially the AAG, NGS, NSF, and NRC—should work together to develop and execute a plan to implement the recommendations in this report.** The committee's recommendations to strengthen the discipline of geography and increase its contributions to science and society are wide ranging in scope and will require a sustained and coordinated effort on the part of these organizations if they are to be realized. By working together, these organizations can leverage their individual efforts to develop a coherent implementation strategy, to monitor and evaluate the long-term effectiveness of the strategy, and to promote effective action by geographers, geographic and related organizations, and policy makers to achieve the long-term objectives of this report.

SUMMARY

If these recommendations are implemented, both science and society will benefit, as will geography itself. Underlying nearly all of the recommendations is the conclusion that the *demand* for contributions from geography and the *supply* capacity, given current resources, are far out of line. Unless significant actions are taken, and taken quickly, either geography's contributions will be severely supply constrained (leading, for instance, to restricted enrollments in university courses and programs) or may decline in quality, as limited professional resources are stretched too thinly.

This conclusion is unavoidable, and it raises questions about the allocation of financial resources. If geography's rediscovered relevance has greater value within science and to society than is currently being realized, the investment of resources should be commensurate with this higher potential. But the issue is not merely one of funding. More importantly, a wide range of institutions and leaders—in government, business, research support, science, education, issue advocacy, the communications media, and geography itself as a discipline—need to raise their levels of awareness of geography's value to science and society and find more effective ways to publicize and utilize geography's perspectives, skills, and knowledge base. Looking toward the next century, realizing geography's potentials will require innovative new partnerships between provider and user, supported and supporter, one science and another, data gatherer and data analyst, and basic research and applications of knowledge.

If geography can be a pathfinder in developing and fulfilling such partnerships, it can survive a difficult transition from scarcity to abundance, and science at large will benefit from many of geography's successes as models for other disciplines. Such a future for the discipline is far from certain, and some of the changing conditions in the 1990s may make it more difficult, but it is worth a concerted effort by all of the interested parties and most of all by geography itself.

References

AAG (Association of American Geographers). 1995. Profiles of the AAG membership 1992–1994. AAG Newsletter 30(4):7.

Abler, R.F., J.S. Adams, and P.R. Gould. 1971. Individual spatial decisions in a descriptive framework. In Spatial Organization: The Geographer's View of the World. Englewood Cliffs, N.J.: Prentice-Hall.

Abler, R.F., M.G. Marcus, and J.M. Olson (eds.). 1992. Geography's Inner Worlds: Pervasive Themes in Contempory American Geography. New Brunswick, N.J.: Rutgers University Press.

Abrahams, A.D., A.J. Parsons, and J. Wainwright. 1995. Effects of vegetation change on interrill runoff and erosion, Walnut Gulch, southern Arizona. Geomorphology 13:37–48.

Adams, J.S. 1991. Housing submarkets in an American metropolis. Pp. 108–126 in Our Changing Cities, J.F. Hart (ed.). Baltimore: Johns Hopkins University Press.

Agnew, J.A. 1987. Place and Politics: The Geographical Mediation of State and Society. Boston: Allen & Unwin.

Agnew, J.A. 1992. Place and politics in post-war Italy: A cultural geography of local identity in the provinces of Lucca and Pistoia. Pp. 52–71 in Inventing Places: Studies in Cultural Geography, Kay Anderson and Fay Gale (eds.). Melbourne: Longman Cheshire.

Agnew, J.A., and S. Corbridge. 1989. The new geopolitics: The dynamics of political disorder. Chapter 10 in A World in Crisis: Geographical Perspectives, R.J. Johnston and P.J. Taylor (eds.). Oxford: Basil Blackwell.

Allen, P., and M. Sanglier. 1979. A dynamic model of growth in a central place system. Geographical Analysis 11:256–273.

Anderson, K., and F. Gale (eds.). 1992. Inventing Places: Studies in Cultural Geography. Melbourne: Longman Cheshire.

Andrews, S.K., and D.W. Tilton. 1993. How multimedia and hypermedia are changing the look of maps. Pp. 348–366 in Proceedings, Auto Carto 11. Bethesda, Md.: American Congress on Surveying and Mapping.

Angel, D.P. 1994. Restructuring for Innovation: The Remaking of the U.S. Semiconductor Industry. New York: Guilford Press.

Anselin, L., R.F. Dodson, and S. Hudak. 1993. Linking GIS and spatial data analysis in practice. Geographical Systems 1:3–23.

Appendini, K., and D.M. Liverman. 1994. Agricultural policy, climate change, and food security in Mexico. Food Policy 19(2):149–163.

Armstrong, M., and R. Marciano. 1995. Massively parallel processing of spatial statistics. International Journal of Geographical Information Systems 9(2):169–189.

Armstrong, M., G. Rushton, and P. Lolonis. 1991. Relationships Between the Birth Weight of Iowa Children and Geographical Accessibility to Obstetrical Care. A report prepared for the Iowa Department of Public Health. Iowa City: Department of Geography, University of Iowa.

Arnfield, A.J. 1982. An approach to the estimation of the surface radiative properties and radiation budgets of cities. Physical Geography 3:97–122.

Arthur, W.B. 1988. Urban Systems and Historical Path Dependence. Cities and Their Vital Systems: Infrastructure Past and Present. Washington, D.C.: National Academy Press.

Atlas of Florida CD-ROM. 1994. Tallahassee: Institute of Science and Public Affairs, Florida State University.

Baker, W.L. 1989a. Landscape ecology and nature reserve design in the Boundary Waters Canoe Area, Minnesota. Ecology 70(1):23–35.

Baker, W.L. 1989b. Macro- and micro-scale influences on riparian vegetation in western Colorado. Annals of the Association of American Geographers 79:65–78.

Bassett, T., and D.E. Crummey (eds.). 1993. Land in African Agrarian Systems. Madison: University of Wisconsin Press.

Batty, M., and P. Longley. 1994. Fractal Cities: A Geometry of Forms and Function. New York: Academic Press.

Bauer, B.O., and J.C. Schmidt. 1993. Waves and sandbar erosion in the Grand Canyon: Applying coastal theory to a fluvial system. Annals of the Association of American Geographers 83:475–495.

Beard, M.K., and B.P. Buttenfield. 1991. NCGIA Research Initiative 7: Visualization of Spatial Data Quality. Technical Paper 91-26. Santa Barbara, Calif.: National Center for Geographic Information and Analysis.

Bebbington, A. 1994. Theory and relevance in indigenous agriculture: Knowledge, agency and organizations. In Rethinking Social Development, D. Booth (ed.). Harlow, U.K.: Longmans.

Bebbington, A., and J. Carney. 1990. Geography in the international agricultural research centers: Theoretical and practical concerns. Annals of the Association of American Geographers 80:34–48.

Bendix, J. 1994. Scale, direction and pattern in riparian vegetation-environment relationships. Annals of the Association of American Geographers 84:652–665.

Berry, B.J.L. 1991. Longwave Rhythms in Economic Development and Political Behavior. Baltimore: Johns Hopkins University Press.

Berry, B.J.L. 1994. The metropolitan frontier: Cities in the modern American-west. Urban Geography 15(8):778–779.

Berry, B.J.L., and J. Parr. 1988. Geography of Market Centers and Retail Distribution, 2nd ed. Englewood Cliffs, N.J.: Prentice-Hall.

Berry, B.J.L., H. Kim, and H.M. Kim. 1994. Innovation diffusion and long waves: Further evidence. Technological Forecasting and Social Change 46(3):289.

Beyers, W.B. 1983. The interregional structure of the U.S. economy. International Regional Science Review 8:213–231.

Blaikie, P., and H.C. Brookfield. 1987. Land Degradation and Society. London: Methuen.

Blaut, J. 1987. Diffusionism: A uniformitarian critique. Annals of the Association of American Geographers 77:30–47.

Bolin, B., B.R. Döös, J. Jäger, and R.A. Warrick (eds.). 1986. The Greenhouse Effect, Climate Change, and Ecosystems. SCOPE 29. Chichester: John Wiley & Sons.

Borchert, J.R. 1967. American metropolitan evolution. Geographical Review 57:301–322.

Borchert, J.R. 1987. America's Northern Heartland. Minneapolis: University of Minnesota Press.

Broecker, W.S. 1994. Massive iceberg discharges as triggers for global climate change. Nature 372:421–424.

Brown, L. 1981. Innovation Diffusion: A New Perspective. London: Methuen.

Brown, L.A., and E.G. Moore. 1971. The intra-urban migration process: A perspective. Pp. 200–209 in Internal Structure of the City, L.S. Bourne (ed.). Toronto: Oxford University Press.

Brunn, S.D. 1995. Geographic research performance based on Annals manuscripts, 1987–1993. The Professional Geographer 47(2).

Brunn, S.D., and T.R. Leinbach (eds.). 1991. Collapsing Space and Time: Geographic Aspects of Communications and Information. London: Harper Collins Academic.

Buttenfield, B.P., and M.K. Beard. 1994. Graphical and geographical components of data quality. Pp. 150–157 in Visualization in Geographic Information Systems, D. Unwin and H. Hearnshaw (eds.). London: Wiley.

Buttenfield, B., and R. McMaster. 1991. Map Generalization: Making Rules for Knowledge Representation. Essex: Longman Group Ltd.

Buttimer, A. 1974. Values in geography. Commission on College Geography Resource Paper No. 24. Washington, D.C.: Association of American Geographers.

Butzer, K.W. 1982. Archaeology as Human Ecology: Theory and Method for a Contextual Approach. Cambridge: Cambridge University Press.

Carney, J. 1993. Converting the wetlands, engendering the environment: The intersection of gender with agrarian change in The Gambia. Economic Geography 69:329–348.

Cartography and Geographic Information Systems. 1995. Special Issue on GIS and Society 22(1).

Casetti, E. 1972. Generating models by expansion method: Applications to geographical research. Geographical Analysis 4:81–91.

Castells, M. 1989. The Informational City. Oxford: Basil Blackwell.

Central Intelligence Agency, Geographic Resources Division. 1995. The Challenge of Ethnic Conflict to National and International Order in the 1990s: Geographical Perspectives. Washington, D.C.: U.S. Government Printing Office.

Cerveny, R.S. 1991. Orbital signals in the diurnal cycle of radiation. Journal of Geophysical Research 96(D9):17,209–17,215.

Chambers, F.B., M.G. Marcus, and L.T. Thompson. 1991. Mass balance of west Gulkana glacier, Alaska. Geographical Review 81:70–86.

Chorley, R.J., and P. Haggett (eds.). 1967. Models in Geography. London: Methuen.

Clark, G. 1985. Judges and the Cities: Interpreting Local Autonomy. Chicago: University of Chicago Press.

Clark, W.A.V. 1992. Comparing cross sectional and longitudinal models of mobility and migration. Environment and Planning A24:1291–1302.

Clark, W.A.V., and P.A. Morrison. 1991. Demographic Paradoxes in the Los Angeles Voting Rights Case. Evaluation Review, vol. 15, 712 pp.

Clark, W.A.V., M.C. Deurloo, and F.M. Dieleman. 1994. Tenure changes in the context of micro level family and macro level economic shifts. Urban Studies 31:137–154.

Cliff, A.D., P. Haggett, J.K. Ord, and G.R. Versey. 1981. Spatial Diffusion: An Historical Geography of Epidemics in an Island Community. New York: Cambridge University Press.

Cliff, A.D., P. Haggett, and J. Ord. 1986. Spatial Aspects of Influenza Epidemics. London: Pion.

Cloke, P., C. Philo, and D. Sadler. 1991. Approaching Human Geography. London: Paul Chapman.

Cohen, S. 1991. Global geo-political change in the post-Cold War era. Annals of the Association of American Geographers 81(4):551–580.

COHMAP (Cooperative Holocene Mapping Project). 1988. Climate changes of the last 18,000 years: Observations and model simulations. Science 241:1043–1052.

Collins, B.M. 1993. Data visualization—Has it all been done before? Pp. 3–28 in Animation and

Scientific Visualization: Tools and Applications, R.A. Earnshaw and D. Watson (eds.). New York: Academic Press.

Conzen, M.P. 1975. The maturing urban system in the United States, 1940–1910. Annals of the Association of American Geographers 67:88–108.

Cook, N.D., and W.G. Lovell (eds.). 1992. Secret Judgments of God: Old World Diseases in Colonial Spanish America. Norman: University of Oklahoma Press.

Cosgrove, D., and S. Daniels (eds.). 1988. The Iconography of Landscape: Essays on the Symbolic Representation, Design, and Use of Past Environments. Cambridge Studies in Historical Geography. Cambridge: Cambridge University Press.

Cowen, D., L. Shirley, P. Noonon, and C. Wiesner. 1995. Toward cadastral-level cartographic analysis using multi-scale spatial data. Pp. 2328–2332 in Proceedings of the International Cartographic Association Congress. Barcelona, Spain.

Currey, D.R. 1994. Semiarid lake basins: Hydrologic patterns. Pp. 405–421 in Geomorphology of Desert Environments, A.D. Abrahams and A.J. Parsons (eds.). London: Chapman and Hall.

Cutter, S.L. 1993. Living with Risk. London: Edward Arnold.

Cutter, S.L., H.L. Renwick, and W.H. Renwick. 1991. Exploitation, Conservation, Preservation: A Geographic Perspective on Natural Resource Use, 2nd ed. New York: John Wiley & Sons.

Dear, M., and J. Wolch. 1987. Landscapes of Despair: From Deinstitutionalization to Homelessness. Princeton, N.J.: Princeton University Press.

Demko, J., and W.B. Wood (eds.). 1994. Reordering the World: Geopolitical Perspectives on the 21st Century. Boulder: Westview Press.

Demeritt, D. 1994. The nature of metaphors in cultural geography and environmental history. Progress in Human Geography 18:163–185.

Dendrinos, D. 1992. The Dynamics of Cities: Ecological Determinism, Dualism, and Chaos. London: Routledge.

Denevan, W.E. 1992. The Native Population of the Americas in 1492, 2nd ed. Madison: University of Wisconsin Press.

DeWispelare, A., L.T. Herren, M. Miklas, and R. Clemen. 1993. Expert Elicitation of Future Climate in the Yucca Mountain Vicinity: Iterative Performance Assessment Phase 2.5. CNWRA 93-016. San Antonio: Center for Nuclear Waste Regulatory Analyses.

Dicken, P. 1992. Global Shift: The Internationalization of Economic Activity, 2nd ed. New York: Guilford Press.

Dieleman, F., and C. Hamnett. 1994. Globalization, regulation and the urban system. Urban Studies 31:357–364.

Dobson, J., and E. Bright. 1991. Coast watch—Detecting change in coastal wetlands. Geo Info Systems (Jan.):36–40.

Dobson, J., R. Ferguson, D. Field, L. Wood, K. Haddad, H. Iredale, V. Klemas, R. Orth, and J. Thomas. 1993. NOAA Coastwatch Change Analysis Project Guidance for Regional Implementation. Coastwatch Change Analysis Project, Coastal Ocean Program, National Oceanic and Atmospheric Administration, U.S. Department of Commerce.

Dorling, D., and S. Openshaw. 1992. Using computer animation to visualize space-time patterns. Environment and Planning B19:639–650.

Downs, R.M. 1994. Being and becoming a geographer: An agenda for geography education. Annals of the Association of American Geographers 84:175–191.

Drysdale, A., and G.H. Blake. 1985. The Middle East and North Africa: A Political Geography. Oxford: Oxford University Press.

Dunn, E.S., Jr. 1980. The Development of the U.S. Urban System. Baltimore: Johns Hopkins University Press.

Earle, C., K. Mathewson, and M.S. Kenzer (eds.). 1996. Concepts in Human Geography. Lanham, Md.: Rowman and Littlefield.

Easterlin, R.A. 1980. Birth and Fortune: The Impact of Numbers on Personal Welfare. New York: Basic Books.
Emel, J.L., and R. Roberts. 1995. Institutional form and its effect on environmental change: The case of groundwater in the southern high plains. Annals of the Association of American Geographers 85(4):664–683.
Erickson, R., and D. Hayward. 1991. The international flows of industrial exports from U.S. regions. Annals of the Association of American Geographers 81(3):371–390.
Eyton, J.R. 1990. Color stereoscopic effect cartography. Cartographica 27(1):20–29.
Fainstein, S., I. Gordon, and M. Harloe. 1992. Divided Cities. Oxford: Basil Blackwell.
Farmer, D. 1990. A rosetta stone for connectionism. Physica D42:153–187.
Feng, Z., W.C. Johnson, Y. Lu, and P.A. Ward. 1994. Climate signals from loess-soil sequences in the central Great Plains, USA. Palaeogeography, Palaeoclimatology, Palaeoecology 110:345–358.
Florida, R., and M. Kenney. 1990. The Breakthrough Illusion. New York: Basic Books.
Forman, R.T., and M. Godron. 1986. Landscape Ecology. New York: John Wiley & Sons.
Gaile, G.L., and C.J. Willmott (eds.). 1989. Geography in America. Columbus, Ohio: Merrill.
Gallup Organization, Inc. 1988. Geography: An International Gallup Survey. Princeton, N.J.
Geography Education Standards Project. 1994. Geography for Life: National Geography Standards 1994. Washington, D.C.: National Geographic Research and Exploration on behalf of the American Geographical Society, the Association of American Geographers, the National Council for Geographic Education, and the National Geographic Society.
Gersmehl, P.J. 1990. Choosing tools: Nine metaphors of four-dimensional cartography. Cartographic Perspectives 5:3–17.
Getis, A., and J.K. Ord. 1992. The analysis of spatial association by the use of distance statistics. Geographical Analysis 24:189–206.
Ghosh, A., and S.L. McLafferty. 1987. Location Strategies for Retail and Service Firms. Lexington, Mass.: D.C. Heath.
Gibbons, J. 1994. Science, Technology, and the Clinton Administration. Paper presented at the annual meeting of the American Association for the Advancement of Science, San Francisco, Feb.
Giddens, A. 1984. The Constitution of Society. Berkeley: University of California Press.
Giddens, A. 1985. The Constitution of Society: Outline of the Theory of Structuration. Berkeley: University of California Press.
Glacken, C.J. 1967. Traces on the Rhodian Shore. Berkeley: University of California Press.
Glasmeier, A.K., and M. Howland. 1995. From Combines to Computers: Rural Services and Development in the Age of Information Technology. Albany: State University of New York Press.
Golding, D., R. Goble, J.X. Kasperson, R.E. Kasperson, J. Seley, G. Thompson, and C. Wolf. 1992. Managing Nuclear Accidents: A Model Emergency Response Plan for Power Plants and Communities. Boulder: Westview Press.
Golding, D., J.X. Kasperson, and R.E. Kasperson. 1994. Preparing for Nuclear Power Plant Accidents: Selected Papers. Boulder: Westview Press.
Golledge, R.G. 1991. Tactual strip maps as navigational aids. Journal of Visual Impairment and Blindness 85(7):296.
Golledge, R., and H. Timmermans. 1988. Behavioral Modelling in Geography and Planning. London: Routledge.
Goodchild, M.F. and D.M. Mark. 1987. The Fractal Nature of Geographic Phenomena. Annals of the Association of American Geographers 77(2):265–278.
Goodchild, M., L. Chih-Chang, and Y. Leung. 1994. Visualizing fuzzy maps. Pp. 158–167 in Visualization in Geographical Information Systems, H. Hearnshaw and D. Unwin (eds.). London: Wiley.
Gould, P. 1989. Geographic dimension of the AIDS epidemic. The Professional Geographer 41(1):71–78.
Gould, P. 1993. The Slow Plague: A Geography of the AIDS Pandemic. Cambridge, Mass.: Blackwell.

Gould, P., and R. Wallace. 1994. Spatial structures and scientific paradoxes in the AIDS pandemic. Geografiska Annaler 76B:105–116.

Gourman, J. 1993. The Gourman Report: A Rating of Graduate and Professional Programs in American and International Universities. Los Angeles: National Education Standards.

Graf, W.I. 1994. Plutonium and the Rio Grande. New York: Oxford University Press.

Gregory, D. 1994. Geographical Imaginations. Cambridge, Mass.: Blackwell.

Gregory, D., and J. Urry (eds.). 1985. Social Relations and Spatial Structures. New York: St. Martins Press.

Grimmond, C.S.B., and T.R. Oke. 1995. Comparison of heat fluxes from summertime observations in the suburbs of four North American cities. Journal of Applied Meteorology 34:873–889.

Grimmond, C.S.B., and C. Souch. 1994. Surface description for urban climate studies: A GIS based methodology. Geocarto International 9:47–59.

Grimmond, C.S.B., C. Souch, and M. Hubble. 1996. The influence of tree cover on summertime energy balance fluxes, San Gabriel Valley, Los Angeles. Climate Research 6(1):45–57.

Grossman, L.S. 1984. Peasants, Subsistence Ecology, and Development in the High Papua New Guinea. Princeton, N.J.: Princeton University Press.

Haag, G., and D. Dendrinos. 1983. Toward a stochastic dynamical theory of location: A nonlinear migration process. Geographical Analysis 15:269–286.

Hägerstrand, T. 1953. Innovationsforloppet ur korologisk synpunkt (Innovation Diffusion as a Spatial Process). Lund, Sweden: C.W.K. Gleerupska.

Hägerstrand, T. 1967. Innovation Diffusion as a Spatial Process. Postscript and Translation by Allan Pred. Chicago: University of Chicago Press. Translated from Innovationsforloppet ur korologisk synpunkt, published in 1953 by C.W.K. Gleerupska, Lund, Sweden.

Hägerstrand, T. 1970. What about people in regional science? Papers of the Regional Science Association 24:7–21.

Haggett, P. 1972. Geography: A Modern Synthesis. New York: Harper & Row.

Haggett, P., A. Cliff, and A. Frey. 1979. Locational Analysis in Human Geography, 2nd ed. London: Edward Arnold.

Hall, S.S. 1992. Mapping the Next Millennium: The Discovery of New Geographies. New York: Random House.

Hall, P., and A. Markusen. 1985. Silicon Landscapes. Boston: Allen Unwin.

Handler, P. 1979. Science, technology, and local achievements. EPRI Journal 4(7):14–19.

Hanson, S. 1986. The Geography of Urban Transportation. New York: Guilford Press.

Harden, C.P. 1991. Andean soil erosion. National Geographic Research & Exploration 7(2):216–231.

Harden, C.P. 1992. Incorporating roads and footpaths in watershed-scale hydrologic and soil erosion models. Physical Geography 13(4):368–385.

Harden, C.P. 1996. Interrelationships between land abandonment and land degradation: A case from the Ecuadorian Andes. Mountain Research and Development 16(3):274–280.

Hardin, G. 1968. The tragedy of the commons. Science 162:1243–1248.

Harley, B. 1990. Cartography, ethics, and social theory. Cartographica 27(12):1–23.

Harley, J.B. 1988. Maps, knowledge, and power. Pp. 277–311 in The Iconography of Landscape: Essays on the Symbolic Representation, Design and Use of Past Environments, D. Cosgrove and S. Daniels (eds.). Cambridge: Cambridge University Press.

Harries, K.D., S.J. Stadler, and R.T. Zdorkowski. 1984. Seasonality and assault: Explorations in inter-neighborhood variation, Dallas, 1980. Annals of the Association of American Geographers 74:590–604.

Harvey, D. 1969. Explanation in Geography. London: Edward Arnold.

Harvey, D. 1973. Social Justice and the City. Baltimore: Johns Hopkins University Press.

Harvey, D. 1982. The Limits to Capital. Oxford: Basil Blackwell.

Harvey, D. 1985a. The Urbanization of Capital. Oxford: Oxford University Press.

Harvey, D. 1985b. Consciousness and the Urban Experience. Oxford: Oxford University Press.

Harvey, D. 1989. The Condition of Postmodernity. Oxford: Blackwell.

Haynes, K.E., and A.S. Fotheringham. 1984. Gravity and Spatial Interaction Models. Beverly Hills, Calif.: Sage Publications.

Hecht, S., and A. Cockburn. 1989. The Fate of the Forest: Developers, Destroyers, and Defenders of the Amazon. London: Verso.

Henderson-Sellers, A. (ed.). 1995. Future Climates of the World: A Modelling Perspective. World Survey of Climatology Vol. 16. Amsterdam: Elsevier Science B.V.

Hirschboeck, K.K. 1991. Climate and floods. Pp. 67–88 in National Water Summary 1988–89—Hydrologic Events and Floods and Droughts. U.S. Geological Survey Water-Supply Paper 2375.

Horn, S.P. 1993. Postglacial vegetation and fire history in the Chirripó páramo of Costa Rica. Quaternary Research 40:107–116.

Huff, D.L. 1963. A probabilistic analysis of shopping center trade areas. Land Economics 39:81–90.

IGBP (International Geosphere-Biosphere Programme). 1994. IGBP in Action: Work Plan 1994–1998. Report No. 28. Stockholm: IGBP.

Isard, W. 1975. Introduction to Regional Science. Englewood Cliffs, N.J.: Prentice-Hall.

Jackson, J.B. 1984. Discovering the Vernacular Landscape. New Haven: Yale University Press.

Jackson, P. (ed.). 1987. Race and Racism: Essays in Social Geography. London: Allen & Unwin.

Jackson, P. 1989. Maps of Meaning. London: Unwin Hyman.

Jackson, P., and J. Penrose (eds.). 1993. Constructions of Race, Place and Nation. Minneapolis: University of Minnesota Press.

James, L.A. 1989. Sustained storage and transport of hydraulic gold mining sediment in the Bear River, California. Annals of the Association of American Geographers 79:570–592.

Janelle, D.G. 1992. The peopling of American geography. Pp. 363–390 in Geography's Inner Worlds: Pervasive Themes in Contemporary American Geography, R.F. Abler, M.G. Marcus, and J.M. Olson (eds.). New Brunswick, N.J.: Rutgers University Press.

Jensen, J., D. Cowen, J. Althausen, and O. Weatherbee. 1993a. An evaluation of Coastwatch change detection protocol in South Carolina. Photogrammetric Engineering and Remote Sensing 59:1039–1046.

Jensen, J., D. Cowen, J. Althausen, S. Narumalani, and O. Weatherbee. 1993b. The detection and prediction of sea level changes on coastal wetlands using satellite imagery and a geographic information system. GeoCarto International 8(4):87–98.

Johnson, J.H., Jr., C. Jones, W. Farrell, and M. Oliver. 1992. The Los Angeles rebellion, 1992: A retrospective view. Economic Development Quarterly 6(4):356–372.

Johnston, R.J., P.J. Taylor, and M.J. Watts (eds.). 1995. Geographies of Global Change: Remapping the World in the Late Twentieth Century. Oxford: Blackwell.

Jones, J.P., and E. Casetti. 1992. Applications of the Expansion Method. London: Routledge.

Jordan, P. 1993. The problems of creating a stable political-territorial structure in hitherto Yugoslavia. Pp. 133–142 in Croatia: A New European State, I. Crkvenci, M. Klemencic, and D. Feletar, eds. Zagreb: Urednici.

Juhl, G. 1994. Wake County develops intelligent parcel management system. Geo Info Systems (June): 44–46.

Kahrl, W.L. (ed.). 1979. The California Water Atlas. Sacramento: State of California.

Kasperson, R.E., and J.M. Stallen. 1991. Communicating Risks to the Public: International Perspectives. Dordrecht: Kluwer Academic Publishers.

Kasperson, J.X., R.E. Kasperson, and B.L. Turner II (eds.). 1995. Regions at Risk: Comparisons of Threatened Environments. Tokyo: United Nations University Press.

Kates, R.W. 1994a. President's column. AAG Newsletter 29(4):1–2.

Kates, R.W. 1994b. Sustaining life on the Earth. Scientific American 271:114–122.

Kates, R.W. 1995. Labnotes from the Jeremiah Experiment: A Hope for a Sustainable Transition. Annals of the Association of American Geographers 85(4):623–640.

Kates, R.W., J.H. Ausubel, and M. Berberian. 1985. Climate Impact Assessment: Studies of the Interaction of Climate and Society. New York: Wiley.

Keith, M., and S. Pile (eds.). 1993. Place and the Politics of Identity. New York: Routledge.

Kelmelis, J.A., D.A. Kirtland, D.A. Nystrom, and N. VanDriel. 1993. From local to global scales: GIS applications at the U.S. Geological Survey. Geo Info Systems 3(9):35–43.

Kesel, R.H., E.G. Yodis, and D.J. McCraw. 1992. An approximation of the sediment budget of the lower Mississippi River prior to major human modification. Earth Surface Processes and Landforms 17:711–722.

Kitzberger, T., T.T. Veblen, and R. Villalba. 1995. Climatic influences on fire regimes along a rainforest-to-xeric woodland gradient in northern Patagonia, Argentina. Unpublished manuscript.

Klemas, V., J. Dobson, R. Ferguson, and K. Haddad. 1993. A coastal land cover classification system for the NOAA Coastwatch Change Analysis Project. Journal of Coastal Research 9(3):862–872.

Kliot, N. 1994. Water Resources and Conflict in the Middle East. New York: Routledge.

Knox, J.C. 1993. Large increases in flood magnitude in response to modest changes in climate. Nature 361:430–432.

Knox, P. 1994. Urbanization. Englewood Cliffs, N.J.: Prentice-Hall.

Krugman, P. 1991. Geography and Trade. Cambridge, Mass.: MIT Press.

Langford, I.H. 1994. Using empirical Bayes estimates in geographical analysis of disease risk. Area 26:142–149.

Lasker, R.D., B.L. Humphreys, and W.R. Braithwaite. 1995. Making a Powerful Connection: The Health of the Public and the National Information Infrastructure. Report of the U.S. Public Health Service, Public Health Policy Coordinating Committee, Bethesda, Maryland, July 6.

Laurini, R., and D. Thomas. 1992. Fundamentals of Spatial Information Systems. New York: Academic Press.

Lee, D. 1990. The status of women in geography: Things change, things remain the same. The Professional Geographer 42:202–211.

Leitner, H., and D. Delaney (guest editors). 1996. Special issue: The Political Construction of Scale. Political Geography.

Lewis, M.W. 1991. Elusive societies: A regional cartographical approach to the study of human relatedness. Annals of the Association of American Geographers 81(4):605–626.

Lewis, M.W. 1992. Wagering the land: Ritual, capital, and environmental degradation in the cordillera of northern Luzon, 1900–1986. Berkeley: University of California Press.

Lindholm, M., and T. Sarjakoski. 1994. Designing a visualization user interface. Pp. 167–184 in Visualization in Modern Cartography, A.M. MacEachren and D.R.F. Taylor (eds.). Oxford: Elsevier Science, Ltd.

Liu, T., and R.I. Dorn. 1996. Understanding spatial variability in environmental changes in drylands with rock varnish microliminations. Annals of the Association of American Geographers 86:187–212.

Liu, K.B., and M.L. Fearn. 1993. Lake-sediment record of late Holocene hurricane activities from coastal Alabama. Geology 21:793–796.

Liverman, D. 1990. Drought impacts in Mexico: Climate, agriculture, technology, and land tenure in Sonora and Puebla. Annals of the Association of American Geographers 80:40–72.

Loveland, T.R., J.W. Merchant, J.F. Brown, D.O. Ohlen, B.C. Reed, P. Olson, and J. Hutchison. 1995. Seasonal land-cover regions of the United States. Annals of the Association of American Geographers 85(2):339–355.

MacEachren, A.M. 1995. How Maps Work: Representation, Visualization, and Design. New York: Guilford Press.

MacEachren, A.M., B.P. Buttenfield, J. Campbell, D. DiBiase, and M. Monmonier. 1992. Visualization. Pp. 99–137 in Geography's Inner Worlds, R.F. Abler, M.G. Marcus, and J.M. Olson (eds.). New Brunswick, N.J.: Rutgers University Press.

MacEachren, A.M., D. Howard, M. von Wyss, D. Askov, and T. Taormino. 1993. Visualizing the health of Chesapeake Bay: An uncertain endeavor. Proceedings, GIS/LIS '93, Minneapolis, MN, 2–4 Nov., 1993, pp. 449–458. Bethesda, Md.: American Society for Photogrammetry and Remote Sensing.

Macmillan, B. 1989. Remodeling Geography. Oxford: Basil Blackwell.

Malanson, G.P. 1993. Riparian Landscapes. Cambridge: Cambridge University Press.

Malanson, G.P., D.R. Butler, and S.J. Walsh. 1990. Chaos theory in physical geography. Physical Geography 11(4):293–304.

Malecki, E.J. 1991. Technology and Development. New York: John Wiley & Sons.

Marcus, W.A., and M.S. Kearney. 1991. Upland and coastal sediment sources in a Chesapeake Bay estuary. Annals of the Association of American Geographers 81:408–424.

Markusen, A.R. 1987. Regions: The Economics and Politics of Territory. Totowa, N.J.: Rowman and Littlefield.

Markusen, A.R., P. Hall, S. Dietrick, and S. Campbell. 1991. The Rise of the Gunbelt: The Military Remapping of Industrial America. New York: Oxford University Press.

Massey, D.B. 1984. Spatial Divisions of Labour: Social Structure and the Geography of Production. London: Methuen.

Mather, J.R., and M. Sanderson. 1996. The Genius of C. Warren Thornthwaite, Climatologist-Geographer. Norman: University of Oklahoma Press.

Mather, J.R., and G.V. Sdasyuk (eds.). 1991. Global Change: Geographical Approaches: A Joint USSR-USA Project Under the Scientific Leadership of Vladimir M. Kotlyakov and Gilbert F. White. Tucson: University of Arizona Press.

McDowell, L. 1993a. Space, place, and gender relations. Part 1: Feminist empiricism and the geography of social relations. Progress in Human Geography 17:157–179.

McDowell, L. 1993b. Space, place, and gender relations. Part 2: Identity, difference, feminist geometries and geographies. Progress in Human Geography 17:305–318.

McDowell, P.F., T. Webb III, and P.J. Bartlein. 1991. Long-term environmental change. Pp. 143–152 in The Earth as Transformed by Human Action, B.L. Turner et al. (eds.). New York: Cambridge University Press.

McMaster, R.B., and K.S. Shea. 1992. Generalization in Digital Cartography. Washington, D.C.: Association of American Geographers.

Medley, K.E. 1993. Primate conservation along the Tana River, Kenya. An examination of forest habitat. Conservation Biology 7:109–121.

Meinig, D.W. 1986 et seq. The Shaping of America (4 volumes planned, two published to date). New Haven, Conn.: Yale University Press.

Meyer, W.B., and B.L. Turner II. 1992. Human population growth and global land-use/cover change. Annual Review of Ecology and Systematics 23:39–61.

Meyer, W.B., and B.L. Turner II (eds.). 1994. Changes in Land Use and Land Cover: A Global Perspective. Cambridge: Cambridge University Press.

Mikesell, M.W., and A.B. Murphy. 1991. A framework for comparative study of minority-group aspirations. Annals of the Association of American Geographers 81(4):581–604.

Mitasova, H., J. Hofierka, M. Zlocha, and R.L. Iverson. 1996. Modeling topographic potential for erosion and deposition using GIS. International Journal of GIS 10(5):629.

Moellering, H. 1989. A practical and efficient approach to the stereoscopic display and manipulation of cartographic objects. Pp. 1–14 in Proceedings, Auto-Carto 9. Baltimore: American Congress of Surveying and Mapping.

Moellering, H., and R. Hogan (eds.). In press. Spatial Database Transfer Standards 2: Characteristics for Assessing Standards and Full Descriptions of the National and International Standards in the World. The ICA Commission on Standards for the Transfer of Spatial Data. London: Elsevier Science.

Moellering, H., and J. Kimerling. 1990. A new digital slope-aspect display process. Cartography and Geographic Information Systems 17(2):151–159.

Moellering, H., and K. Wortman. 1994. Technical Characteristics for Assessing Standards for the Transfer of Spatial Data. In The ICA Commission on Standards for the Transfer of Spatial Data. Bethesda, Md.: American Congress on Surveying and Mapping.

Monmonier, M. 1989. Geographic brushing: Enhancing exploratory analysis of the scatterplot matrix. Geographical Analysis 21(1):81–84.

Monmonier, M. 1992. Authoring graphics scripts: Experiences and principles. Cartography and Geographic Information Systems 19(4):247–260.

Morrill, R.L. 1981. Political Redistricting and Geographic Theory. Resource Publications in Geography. Washington, D.C.: Association of American Geographers.

Morrison, P.A., and W.A.V. Clark. 1992. Local redistricting: The demographic context of boundary drawing. National Civic Review 81(1):57–63.

Mortimore, M.J. 1989. Adapting to Drought. Cambridge: Cambridge University Press.

Mossa, J., and W.J. Autin. 1996. Geographic and geologic aspects of aggregate production in Louisiana. In Aggregate Resources: A Global Perspective, P. Bobrowski (ed.). Rotterdam: A. A. Balkema.

Murphy, A.B. 1989. Territorial policies in multiethnic states. The Geographical Review 79(4):410–421.

Murphy, A.B. 1993. Emerging regional linkages within the European community: Challenging the dominance of the state. Tijdschrift voor Economische en Sociale Geografie 84(2):103–118.

Murphy, A.B. 1995a. Economic regionalization and Pacific Asia. Geographical Review 85(2):127–140.

Murphy, A.B. (rapporteur). 1995b. Geographic Approaches to Democratization. Report to the National Science Foundation, Division of Social, Behavioral, and Economic Research, Arlington, Va.

National Center for Education Statistics. 1993. Digest of Education Statistics, 1993. Washington, D.C.: U.S. Government Printing Office.

NRC (National Research Council), Committee on Geography. 1965. The Science of Geography. Washington, D.C.: National Academy Press.

NRC (National Research Council), Committee on the Human Dimensions of Global Change. 1992a. Global Environmental Change: Understanding the Human Dimensions. Washington, D.C.: National Academy Press.

NRC (National Research Council), Panel on the Policy Implications of Greenhouse Warming. 1992b. Implications of Greenhouse Warming: Mitigation, Adaptation, and the Science Base. Washington, D.C.: National Academy Press.

NRC (National Research Council), Committee on Science, Engineering, and Public Policy. 1993a. Science, Technology, and the Federal Government. Washington, D.C.: National Academy Press.

NRC (National Research Council), Mapping Science Committee. 1993b. Toward a Coordinated Spatial Data Infrastructure for the Nation. Washington, D.C.: National Academy Press.

NRC (National Research Council), Committee for the Human Dimensions of Global Change. 1994. Science Priorities for the Human Dimensions of Global Change. Washington, D.C.: National Academy Press.

NRC (National Research Council), Committee for the Study of Research-Doctorate Programs in the United States. 1995. Research Doctorate Programs in the United States. Washington, D.C.: National Academy Press.

NSF (National Science Foundation). 1994. Guide to Programs, Fiscal Year 1995. NSF 94-91. NSF: Arlington, Va.

Oke, T.R. 1979. Review of Urban Climatology 1973–1976. Technical Note No. 169. Geneva: World Meteorological Organization.

Oke, T.R. 1987. Boundary Layer Climates, Second Edition. New York: Methuen.

Openshaw, S. 1995. Developing automated and smart spatial pattern exploration tools for geographical information systems applications. The Statistician 44(1):3–16.

Openshaw, S., M. Charlton, C. Wymer, and A.W. Craft. 1987. A Mark I Geographical Analysis

Machine for the automated analysis of point data sets. International Journal of Geographical Information Systems 1:335–358.
Openshaw, S., M. Charlton, A.W. Craft, and J.M. Birch. 1988. Investigations of leukemia clusters by the use of a geographical analysis machine. Lancet I:272–273.
OSTP (Office of Science and Technology Policy). 1994. Science in the National Interest. Washington, D.C.: U.S. Government Printing Office.
Palm, R. 1990. Natural Hazards: An Integrated Framework for Research and Planning. Baltimore: Johns Hopkins University Press.
Parry, M.L. 1990. Climate Change and World Agriculture. London: Earthscan Publications Ltd.
Parry, M.L., T.R. Carter, and N.T. Konijn. 1988. The Impact of Climate Variation on Agriculture. Dordrect: Kluwer.
Peet, R. (ed.). 1987. International Capitalism and Industrial Restructuring. London: Allen & Unwin.
Peet, R., and N. Thrift (eds.). 1989. New Models in Geography. London: Unwin Hyman.
Peuquet, D.J. 1994. It's about time: A conceptual framework for the representation of temporal dynamics in geographic information systems. Annals of the Association of American Geographers 84:441–461.
Peuquet, D.J. 1988. Representations of geographic space: Toward a conceptual synthesis. Annals of the Association of American Geographers 78(3):373–394.
Pickles, J. (ed.). 1995a. Ground Truth: The Social Implications of Geographic Information Systems. New York: Guilford Press.
Pickles, J. 1995b. Representations in an electronic age: Geography, GIS, and democracy. In Ground Truth, J. Pickles (ed.). New York: Guilford Press.
Pike, R.J., and G.P. Thelin. 1989. Shaded relief map of U.S. topography from digital elevations. EOS: Transactions of the American Geophysical Union 70(38):cover, 843, 853.
Pines, D. (ed.). 1986. Emerging Syntheses in Science. Redwood City, Calif.: Addison Wesley.
Plane, D.A. 1993. Demographic influences on migration. Regional Studies 27:375–383.
Plane, D.A., and P.A. Rogerson. 1991. Tracking the baby boom, the baby bust, and the echo generations: How age composition regulates U.S. migration. The Professional Geographer 43:416–430.
Powers, A., P. Wright, M. Pucherelli, and D. Wegner. 1994. GIS efforts target long-term resource monitoring. GIS World 7(5):36–39.
Pred, A. 1977. City-Systems in Advanced Economies. New York: John Wiley.
Pred, A. 1981. Urban Growth and City-System Development in the United States, 1840–1860. Cambridge, Mass.: Harvard University Press.
Prentice, I.C., P.J. Bartlein, and T. Webb III. 1991. Vegetation and climate change in eastern North America since the last glacial maximum. Ecology 72(6):2038–2056.
Pulido, L. 1996. Environmentalism and Economic Justice. Two Struggles in the Southwest. Tucson: University of Arizona Press.
Ralston, B. 1994. Object oriented spatial analysis. Pp. 165–185 in Spatial Analysis and GIS, S. Fortheringham and P. Rogerson (eds.). London: Taylor & Francis, Ltd.
Reuss, M. 1993. Water Resources People and Issues: Interview with Gilbert F. White. Publication EP-870-1-43. Fort Belvoir, Va.: U.S. Army Corps of Engineers.
Riebsame, W., W.B. Meyer, and B.L. Turner II. 1994. Modeling land use/cover as part of global environmental change. Climatic Change 28(1):45–64.
Roberts, R.S., and J. Emel. 1992. Uneven development and the tragedy of the commons: Competing images for nature-society analysis. Economic Geography 68:249–271.
Root, T.L., and S.H. Schneider. 1995. Ecology and climate: Resource strategies and implications. Science 269:334–341.
Rose, H.M. 1971. The Black Ghetto: A Spatial Behavioral Perspective. New York: McGraw-Hill.
Rose, G. 1993. Feminism and Geography: The Limits of Geographical Knowledge. Minneapolis: University of Minnesota Press.

Rosenzweig, C., M.L. Parry, and G. Fischer. 1995. World food supply. Pp. 27–56 in As Climate Changes: International Impacts and Implications, K.M. Strzepek and J.B. Smith (eds.). Cambridge: Cambridge University Press.
Ruggie, J.G. 1993. Territoriality and beyond: Problematizing modernity in international relations. International Organization 41:294–303.
Rundstrom, R.A., and M.S. Kenzer. 1989. The decline of fieldwork in human geography. The Professional Geographer 41:294–303.
Rushton, G. 1988. The Roepke lecture in economic geography: Location theory, location-allocation models and service development planning in the Third World. Economic Geography 64(2):97–120.
Sack, R.D. 1981. Territorial bases of power. Pp. 53–71 in Political Studies from Spatial Perspectives, A.D. Burnett and P.J. Taylor (eds.). New York: John Wiley & Sons.
Sack, R.D. 1986. Human Territoriality: Its Theory and History. Cambridge: Cambridge University Press.
Sassen, S. 1991. The Global City: New York, London, and Tokyo. Princeton, N.J.: Princeton University Press.
SAST (Scientific Assessment and Strategy Team). 1993. Science for Floodplain Management into the Twenty-first Century. Preliminary Report of the Scientific Assessment and Strategy Team, Report of the Interagency Floodplain Management Review Committee to the Administration Floodplain Management Task Force. Washington, D.C.: U.S. Government Printing Office.
Sauer, J.D. 1988. Plant Migration: The Dynamics of Geographic Patterning in Seed Plant Species. Berkeley: University of California Press.
Savage, M. 1993. Ecological disturbance and nature tourism. Geographical Review 83(3):290–300.
Saxenian, A. 1994. Regional Advantage: Culture and Competition in Silicon Valley and Route 128. Cambridge, Mass.: Harvard University Press.
Sayer, A. 1993. Method in Social Science: A Realist Approach, 2nd ed. London: Routledge.
Schmidt, J.C. 1990. Recirculating flow and sedimentation in the Colorado River in Grand Canyon, Arizona. Journal of Geology 98:709–724.
Schroeder, R. 1993. Shady practice: Gender and the political ecology of resource stabilization. Economic Geography 69(4):349–365.
Schoenberger, E. 1988. Multinational corporations and the new international division of labor: A critical appraisal. International Regional Science Review 11:105–119.
Scott, A.J. 1988a. Metropolis: From the Division of Labor to Urban Form. Berkeley: University of California Press.
Scott, A.J. 1988b. Flexible production systems and regional development: The rise of new industrial spaces in North America and western Europe. International Journal of Urban and Regional Research 12:171–186.
Scott, A.J. 1988c. New Industrial Spaces. London: Pion.
Scott, A.J. 1993. Technopolis: High-Technology Industry and Regional Development in Southern California. Berkeley: University of California Press.
Scuderi, L.A. 1987. Late-Holocene upper timberline variations in the southern Sierra Nevada. Nature 325:242–244.
Sheppard, E. 1985. Urban system population dynamics: Incorporating nonlinearities. Geographical Analysis 17:47–73.
Sheppard, E., and T.J. Barnes. 1990. The Capitalist Space Economy: Analytical Foundations. London: Unwin & Hyman.
Sheppard, E., R.P. Haining, and P. Plummer. 1992. Spatial pricing in interdependent markets. Journal of Regional Science 32:55–75.
Shiffer, M. 1993. Augmenting geographic information with collaborative multimedia technologies. Pp. 367–375 in Proceedings, Auto-Carto 11. Bethesda, Md.: American Congress on Surveying and Mapping.

Shrestha, N., and D. Davis, Jr. 1989. Minorities in geography: Some disturbing facts and policy measures. The Professional Geographer 41:410–421.

Sloggett, G., and C. Dickason. 1986. Ground-Water Mining in the United States. Economic Research Service, U.S. Department of Agriculture. Washington, D.C.: U.S. Government Printing Office.

Smith, N. 1992. Geography, difference and the politics of scale. Pp. 57–79 in Postmodernism and the Social Sciences, J. Doherty, E. Graham, and M. Malek (eds.). London: Macmillan.

Smith, M.P., and J. Feagin (eds.). 1987. The Capitalist City: Global Restructuring and Community Politics. Oxford: Basil Blackwell.

Smith, T.R., W.A.V. Clark, J.O. Huff, and P. Shapiro. 1979. A decision making and search model for intraurban migration. Geographical Analysis 11:1–22.

Soja, E.W. 1989. Postmodern Geographies: The Reassertion of Space in Critical Social Theory. London and New York: Verso.

Soule. M.E. 1991. Conservation: Tactics for a constant crisis. Science 253:744–750.

Storper, M., and R. Walker. 1989. The Capitalist Imperative: Territory, Technology, and Industrial Growth. Oxford: Basil Blackwell.

Strezepek, K., and J. Smith (ed.). 1979. If Climate Should Change: International Effects of Global Warming. Cambridge: Cambridge University Press.

Taaffe, E. 1993. Spatial analysis: Development and outlook. Urban Geography 14(5):422–433.

Taaffe, E.J., I. Burton, N. Ginsburg, P.R. Gould, F. Lukermann, and P.L. Wagner (panelists). 1970. Geography. Englewood Cliffs, N.J.: Prentice-Hall. 143 pp.

Taylor, P.J. 1993. Political Geography: World-Economy, Nation-State, and Locality, 3rd ed. Harlow, England: Longman Scientific and Technical.

Terjung, W.H. 1982. Process-Response Systems in Physical Geography. Bonn: Ferd. Dummlers Verlag.

Terjung, W.H., and P.A. O'Rourke. 1980. Simulating the causal elements of urban heat islands. Boundary Layer Meteorology 19:93–118.

Terjung, W.H., S.S.-F. Louis, and P.A. O'Rourke. 1976. Toward an energy budget model of photosynthesis predicting world productivity. Vegetatio 32:31–53.

Thomas, W.T., Jr. (ed.). 1956. Man's Role in Changing the Face of the Earth. Chicago: University of Chicago Press.

Thompson, R.S., C. Whitlock, S.P. Harrison, W.G. Spaulding, and P.J. Bartlein. 1993. Vegetation, lake-levels and climate in the western United States. Pp. 468–513 in Global Climates Since the Last Glacial Maximum, H.E. Wright, Jr. et al. (eds.). Minneapolis: University of Minnesota Press.

Thornthwaite, C.W. 1953. Operations Research in Agriculture. Journal of the Operations Research Society of America 1(2):33–38.

Thornthwaite, C.W., and J.R. Mather. 1955. The water balance. Publications in Climatology 8:1–104.

Thrift, N. 1989. The Geography of International Economic Disorder. Pp. 12–67 in A World in Crisis?, R.J. Johnston and P.J. Taylor (eds.). London: Basil Blackwell.

Tobler, W. 1969. Geographical filters and their inverses. Geographical Analysis 1:234–253.

Tobler, W. 1981. Depicting federal fiscal transfers. The Professional Geographer 33(4):419–422.

Tobler, W., U. Deichmann, J. Gottsegen, and K. Maloy. 1995. The Global Demography Project.Technical Report 95–6. Santa Barbara, Calif.: National Council for Geographic Information and Analysis.

Townshend, J.R.G. 1992. Improved Global Data for Land Applications: A Proposal for a New High Resolution Data Set. Report of the Land Cover Working Group of IGBP-DIS. IGBP Report #20. Stockholm.

Trimble, S.W., F.W. Weirich, and B.L. Hoag. 1987. Reforestation and the reduction of water yield on the southern Piedmont since circa 1940. Water Resources Research 23:425–437.

Tuan, Y.-F. 1974. Topophilia: A Study of Environmental Perception, Attitudes, and Values. Englewood Cliffs, N.J.: Prentice-Hall.

Tuan, Y.-F. 1976. Humanistic geography. Annals of the Association of American Geographers 66:266–276.

Turner, B.L., II, and D. Varlyguin. 1995. Foreign-area expertise in U.S. geography: An assessment of capacity based on foreign-area dissertations, 1977–1991. The Professional Geographer 47(3):308–314.

Turner, B.L., II, W.C. Clark, R.W. Kates, J.F. Richards, J.T. Mathews, and W.B. Meyer. 1990. The Earth as Transformed by Human Action: Global and Regional Changes in the Biosphere Over the Past 300 Years. Cambridge: Cambridge University Press.

Turner, B.L., II, D. Skole, S. Sanderson, G. Fischer, L. Fresco, and R. Leemans. 1995. Land-Use and Land-Cover Change: Science Research Plan. IGBP Report #35/HDP Report #7. Stockholm and Geneva.

Urban Geography. 1991. Special Issue on the Urban Underclass. Rose, H.M. (ed.). 12(6).

U.S. Department of Education. 1992. America 2000: An Education Strategy. Washington D.C.: U.S. Department of Education.

U.S. Department of Labor. 1991. What Work Requires of Schools: A SCANS Report for America 2000. Washington, D.C.: U.S. Department of Labor.

USGCRP (U.S. Global Change Research Program). 1994. Our Changing Planet: The FY 1995 U.S. Global Change Research Program. A Supplement to the President's Fiscal Budget Year Budget. Washington, D.C.: Coordinating Office of the U.S. Global Change Research Program.

USGS (U.S. Geological Survey). 1992. Geographic Information Systems. Washington, D.C.: USGS.

Vale, T.R. 1982. Plants and People: Vegetation Change in North America. Washington, D.C.: Association of American Geographers.

Veblen, T.T., and D.C. Lorenz. 1988. Recent vegetation changes along the forest/steppe ecotone of northern Patagonia. Annals of the Association of American Geographers 78:93–111.

Veblen, T.T., M. Mermoz, C. Martin, and E. Ramilo. 1989. Effects of exotic deer on forest structure and composition in northern Patagonia. Journal of Applied Ecology 26:711–724.

Veblen, T.T., T. Kitzberger, and A. Lara. 1992. Disturbance and forest dynamics along a transect from Andean rain forest to Patagonian shrublands. Journal of Vegetation Science 3:507–520.

Wallin, T., and C. Harden. 1996. Quantifying trail-related soil erosion at two sites in the humid tropics: Jatun Sacha, Ecuador, and La Selva, Costa Rica. Ambio XXV(7): 517–522.

Ward, D. 1971. Cities and Immigrants: A Geography of Change in Nineteenth-Century America. New York: Oxford University Press.

Warf, B., and J. Cox. 1993. The U.S.-Canada free trade agreement and commodity transportation services among U.S. states. Growth and Change 24:341–354.

Watts, M.J. 1983. Silent Violence: Food, Famine and Peasantry in Northern Nigeria. Berkeley: University of California Press.

Webb, T., III, and P.J. Bartlein. 1992. Global changes during the past 3 million years: Climatic controls and biotic responses. Annual Review of Ecology and Systematics 23:141–173.

Webber, M.J. 1987. Rates of profit and interregional flows of capital. Annals of the Association of American Geographers 77:63–75.

Weiss, C.H. (ed.). 1977. Using Social Research in Public Policy Making. Lexington, Mass.: Heath.

Whitlock, C. 1993. Postglacial vegetation and climate of Grand Teton and Southern Yellowstone national parks. Ecological Monographs 63:173–198.

Whitmore, T.M. 1992. Disease and Death in Early Colonial Mexico: Simulating Amerindian Depopulation. Dellplain Latin American Geography Series. Boulder: Westview Press.

Wilbanks, T.J. 1985. Geography and public policy at the national scale. Annals of the American Association of Geographers 75:4–10.

Wilbanks, T.J. 1994. Sustainable development in geographic context. Annals of the Association of American Geographers 84:541–557.

Wilbanks, T.J., and R. Lee. 1985. Policy analysis in theory and practice. Pp. 273–303 in Large-

Scale Energy Projects: Assessment of Regional Consequences, T.R. Lakshmanan and B. Johansson (eds.). Amsterdam: North Holland.
Williamson, P. 1992. Global Change: Reducing Uncertainties. International Geosphere-Biosphere Programme. Stockholm: Royal Swedish Academy of Sciences.
Willmott, C.J., and D.R. Legates. 1991. Rising estimates of terrestrial and global precipitation. Climate Research 1:179–186.
Willmott, C.J., S.M. Robeson, and J.J. Feddema. 1994. Estimating continental and terrestrial precipitation averages from rain-gauge networks. International Journal of Climatology 14:403–414.
Wilm, H.G., C.W. Thornthwaite, E.A. Colman, N.W. Cummings, A.R. Croft, H.T. Gisborne, S.T. Harding, A.H. Hendrickson, M.D. Hoover, I.E. Houk, J. Kittredge, C.H. Lee, C.G. Rossby, T. Saville, and C.A. Taylor. 1944. Report of the Committee on Transpiration and Evaporation, 1943–44. Transactions, American Geophysical Union 25:683–693.
Wilson, A., J. Coelho, S. Macgill, and H. Williams. 1981. Optimization in Locational and Transport Analysis. New York: John Wiley & Sons.
Wolch, J., and M. Dear (eds.). 1989. The Power of Geography: How Territory Shapes Social Life. Boston: Unwin Hyman.
Wolch, J., and M. Dear. 1993. Malign Neglect: Homelessness in an American City. San Francisco: Jossey-Bass Publishers.
Wolpert, J. 1965. Behavioral aspects of the decision to migrate. Papers of the Regional Science Association 15:159–169.
Wood, D. 1992. The Power of Maps. New York: Guilford Press.
Wood, W.B. 1994. Forced migration: Local conflicts and international dilemmas. Annals of the Association of American Geographers 84(4):607–634.
Wright, H.E., Jr., and P.J. Bartlein. 1993. Reflections on COHMAP. Holocene 3:89–92.
Wright, H.E., Jr., J.E. Kutzbach, T. Webb III, W.F. Ruddiman, F.A. Street-Perrott, and P.J. Bartlein (eds.). 1993. Global Climates Since the Last Glacial Maximum. Minneapolis: University of Minnesota Press.
Yapa, L. 1989. Low-cost map overlay analysis using computer-aided design. Environmental Planning B 16:377–391.
Young, K.R. 1992. Biogeography of the montane forest zone of the eastern slopes of Peru. Pp. 119–140 in Biogeografía, Ecología, y Conservación del Bosque Montano en el Peru, K.R. Young and N. Valencia (eds.). Memorias del Museo de Historia Natural 21. Lima, Peru: Universidad Nacional Mayor de San Marcos.
Zimmerer, K.S. 1991. Wetland production and smallholder persistence: Agricultural change in a highland Peruvian region. Annals of the Association of American Geographers 81(3):443–463.
Zimmerer, K.S. 1994. Human geography and the ''new ecology'': The prospect and promise of integration. Annals of the Association of American Geographers 84(1):108–125.

Appendix A

Enrollment and Employment Trends in Geography

The Association of American Geographer's (AAG's) Employment Forecasting Committee[1] prepared the following analysis of present and future enrollment and employment trends for geographers at the request of the Rediscovering Geography Committee. The Employment Forecasting Committee's analysis was published in the August 1995 issue of the *Professional Geographer* and is reprinted here with the permission of the AAG and Blackwell Publishers.

[1]Patricia Gober (Chair), Amy K. Glasmeier, James M. Goodman, David A. Plane, Howard A. Stafford, and Joseph S. Wood.

Employment Trends in Geography, Part 1: Enrollment and Degree Patterns*

Patricia Gober
Arizona State University

Amy K. Glasmeier
Pennsylvania State University

James M. Goodman
National Geographic Society

David A. Plane
University of Arizona

Howard A. Stafford
University of Cincinnati

Joseph S. Wood
George Mason University

This paper is the first in a series of three papers dealing with the current and future labor market for geographers. It is based on a report prepared by the Association of American Geographers' (AAG) Employment Forecasting Committee to the National Research Council's (NRC) Rediscovering Geography Committee. This report provides a data-based analysis of the past and future supply of geographers, the current labor market conditions in the field, and the factors likely to influence the future demand for geographers (faculty hiring, geographic education initiatives, trends in private-sector jobs, etc.).

Each year some 4,000 individuals receive degrees in geography from America's institutions of higher education. They, or some portion of them, make up the new supply of geographers entering the labor market. In the near future (up to five years), the availability of new geographers is related to the number of geography students now in the educational pipeline. Their current specialties, and the specialties of the programs from which they come, tell us about the types of skills and the kinds of interests to be held by future labor force entrants. In the longer term (five to ten years), the number of new geographers will be influenced by geographic education initiatives at the precollegiate level. More and better geographic instruction in elementary and secondary schools will expose more students to geography as a field of study and as a potential career path. The purposes of this paper are to (1) review degree and enrollment trends in geography, (2) assess the "trickle-up" effects of geographic education initiatives at the precollegiate level, and (3) investigate the characteristics of future supply as evidenced by the types of occupations for which geography departments are now preparing students.

Background

Previous attempts to examine employment trends in geography focused on the academic job market (Hart 1966, 1972; Hausladen and Wyckoff 1985; Suckling 1994; Miyares and McGlade 1994) or relied exclusively on AAG membership data (Goodchild and Janelle 1988; Janelle 1992). Early "manpower" studies by John Fraser Hart matched predictions of the future supply of new doctorates in geography with estimates of new teaching jobs in colleges and universities and developed scenarios of surplus and deficit in the academic labor market. Hausladen and Wyckoff examined age profiles of topical and areal specialties from the 1982 *AAG Directory* with an eye toward predicting the effects of future retirements on the field. More recent investigations by Suckling and by Miyares and McGlade focused on the demand side of the employment equation and examined changes in the number, rank, location, and specialties of jobs advertised in *Jobs in Geography*.

Michael Goodchild and Donald Janelle analyzed trends in the intellectual structure of the discipline as manifest in AAG specialty group membership and topical proficiencies. Goodchild and Janelle's findings provided valuable insights into changes in the nature of geographic thought and training. Technical expertise and interest in geographic information systems (GIS) were burgeoning, especially among young geographers, while regionally oriented

*This article is excerpted from the appendix of the report of the National Research Council's Rediscovering Geography Committee. The authors wish to thank John Harner, Mark Patterson, and Will Mitchell for their technical support, Barbara Trapido for figure preparation, and the NRC Rediscovering Geography Committee for their many helpful suggestions. Special thanks go to Kevin Crowley, Program Officer at the National Research Council, for his constructive comments, for serving as our liaison with the NRC Committee, and for his moral support of this research.

Professional Geographer, 47(3) 1995, pages 317–328 © Copyright 1995 by Association of American Geographers.
Initial submission, November 1994; final acceptance, January 1995.
Published by Blackwell Publishers, 238 Main Street, Cambridge, MA 02142, and 108 Cowley Road, Oxford, OX4 1JF, UK.

specialties were shrinking. Reliance on AAG membership data, however, limits the applicability of these studies to college and university professors and their students—individuals who make up about two-thirds of the AAG's membership but a relatively small share of the total number of new geographers entering the labor market.

Degree and Enrollment Trends

The National Center for Education Statistic's (NCES) annual *Digest of Education Statistics* is the most accurate guide to the number of degrees granted in geography in the United States. Published continuously since 1948–1949, these data are compiled from questionnaires returned by each institution of higher education in the United States. Geography is listed as a separate field within the larger category of the "social sciences." Although not all individuals who receive degrees in geography directly enter the labor market for geographic skills (some go on to graduate school and others seek employment in occupations unrelated to geography), this series is the most sensitive barometer available of the supply of new geographers.

The NCES series reveals a surge in geography degrees granted in the late 1960s and early 1970s as the first wave of the baby boom generation began to graduate from colleges and universities, as a higher proportion of young people sought access to higher education, and as amendments to the National Defense Education Act (NDEA) in 1964 bolstered graduate enrollments (Fig. 1). Possible factors specific to geography include curricular reforms in the form of the High School Geography Project and Commission on College Geography and emerging technological advances in remote sensing and quantitative techniques that created opportunities for geographers outside of education. The number of bachelor's degrees peaked in 1971–1972 at 4,300, the number of master's one year later at almost 800, and Ph.D.s one year after that at 217.

From their high point in the early 1970s, the number of geography degrees declined slowly but steadily. At their nadir in 1987–1988, the number of new geographers entering the labor market was just two-thirds what it had been 15 years earlier. This downward pattern was not due to trends in higher education as a whole, as the total numbers of bachelor's, master's, and doctoral degrees in the United States grew steadily during the late 1970s and early 1980s,

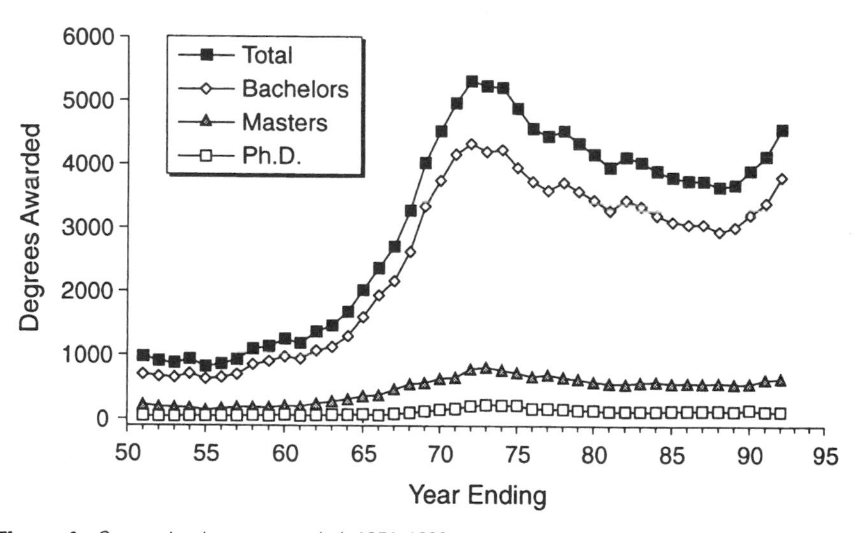

Figure 1: *Geography degrees awarded, 1951–1992.*

despite declines in the traditional college-age population (NCES 1993a).

A turnaround in the number of geography degrees conferred at the undergraduate and masters levels began in 1988. Subsequently, the discipline has outpaced higher education as a whole in the increase in bachelor's and master's degrees granted. Between 1988–1989 and 1990–1991, the number of bachelor's degrees in geography grew by 13% compared with 7% for all fields. The number of master's degrees in geography grew at a rate of 14% compared with 9% for all fields. New Ph.D. holders in geography remained steady while the number of doctorates overall increased by 10% (NCES 1993a). The most recent figures for 1991–1992 show explosive growth at the bachelor's level and continued growth at the master's. Several factors are probably responsible for the renewed vigor in geography's degree production at the bachelor's and master's levels. They include rising public concern with environmental and international problems, greater attention to geographic education at the precollegiate and collegiate levels, and technological advances in GIS that provide new geographers with highly marketable skills.

Increases in the number of geography degrees are broadly consistent with trends in the social sciences, where the number of degrees fell consistently and dramatically from the early 1970s through the mid-1980s (Fig. 2). After 1987, degrees conferred in the social sciences in addition to geography, including anthropology, economics, political science, and sociology, rose sharply. Recent growth in the social sciences has been linked to the shift away from business majors, rising concern over the environment and crime, the growing elderly and homeless populations, and the increasingly competitive global economy (Bureau of Labor Statistics 1994, 120). The increased presence of women and minorities on university campuses also has been associated with the growth in social science fields (NCES 1993a). In the late 1980s female-dominated fields like education, psychology, and the social sciences experienced faster-than-average growth while slower-than-average growth occurred in traditionally male-dominated areas such as business

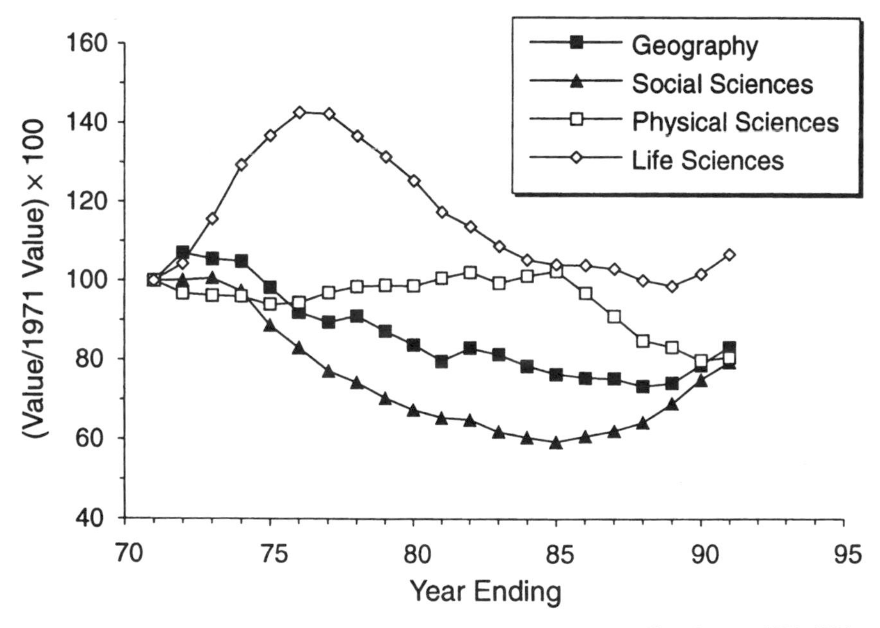

Figure 2: *Degrees in geography, social sciences, physical sciences, and life sciences, 1971–1991.*

and management, engineering, and the physical sciences.

The physical sciences, including the closely related fields of geology and meteorology in addition to astronomy, chemistry, and physics, experienced steady degree production throughout the 1970s and early 1980s. The number of graduates sharply declined after 1985, especially at the bachelor's level (Fig. 2). The decline in new degree holders has been attributed, in part, to the fact that students are not obtaining in the high schools the demanding math and science background that is necessary for pursuing a university degree in the physical sciences (National Science Board 1993). The life sciences grew rapidly during the early and mid-1970s, declined steadily during the 1980s, and then turned upward in the late 1980s, at about the same time that the number of degrees in geography began to rise.

Geography's pattern of degree production during the past 20 years reflects, in part, the patterns of its closely related fields. Geography's early losses were similar to, but not as severe as, those for the social sciences as a whole. The discipline's upturn during the late 1980s is mirrored in both the social and life sciences categories. Ironically, the experience of the physical sciences correlates with geography's, but in an inverse fashion. The physical sciences grew slightly during the late 1970s and early 1980s at the same time that geography declined. They reached their peak at about the time that geography reached its low point in the late 1980s. In recent years, the number of geography degrees has increased while the number of physical science degrees has declined.

Enrollments in geography programs provide another picture of labor supply conditions in the field, although the connection between supply and enrollment is less direct than the association between supply and degrees granted. Enrollment figures are unreliable, particularly at the undergraduate level. Some students are forced by institutional rules to declare their majors before they have settled on a career path. Others change majors, but these changes are not reflected in institutional accounting systems. As a result, enrollment figures are best seen as a general guide to current and future trends in the field.

The most complete record of graduate enrollments in geography programs comes from the NSF (1993). These data reveal that graduate enrollments in geography dipped about 10% from 1981 through 1985 but rose steadily thereafter (Fig. 3). This pattern fits with trends in geography graduate degrees, which bottomed out three years later in 1988 but are now on the rise. It also suggests that the upturn in master's degrees, from 555 in 1990 to 622 in 1991 and 639 in 1992, is not an aberration but the harvest of healthy enrollment growth during the late 1980s.

Geography's pattern is roughly parallel to that of the social science category, although the discipline's early-decade decline in enrollments was less severe and its late-decade rebound was more pronounced. Predictably, trends for the social sciences are smoother because they even out the ebbs and flows of individual disciplines. Graduate enrollments in the environmental sciences, including the atmospheric sciences, the geosciences, and oceanography, rose steadily until mid-decade, declined during the late 1980s, and began to climb again after 1989.

Geography's late-decade burst in enrollments was unmatched by individual fields in the social and environmental sciences. From their low point in 1985, graduate enrollments in geography grew by almost one-third by 1991 (Table 1). Student interest in sociology, anthropology, and political science was strong but not equal to the rise in geography. Enrollments in economics changed little. The poor performance of the category of environmental sciences between 1985 and 1991 was primarily the result of the dismal performance of the geosciences (geology), which comprises more than one-half of the enrollment in the environmental sciences category. Declines in the domestic oil and gas industry and mineral exploration drastically reduced the demand for geologists. The U.S. Geological Survey, the major government employer of geologists, and individual state surveys also downsized during this period.

Recent increases in graduate geography enrollments should translate into an increasing supply of highly trained geographers entering the labor market in the near future. Indeed, because 1991 is the most recent year for which

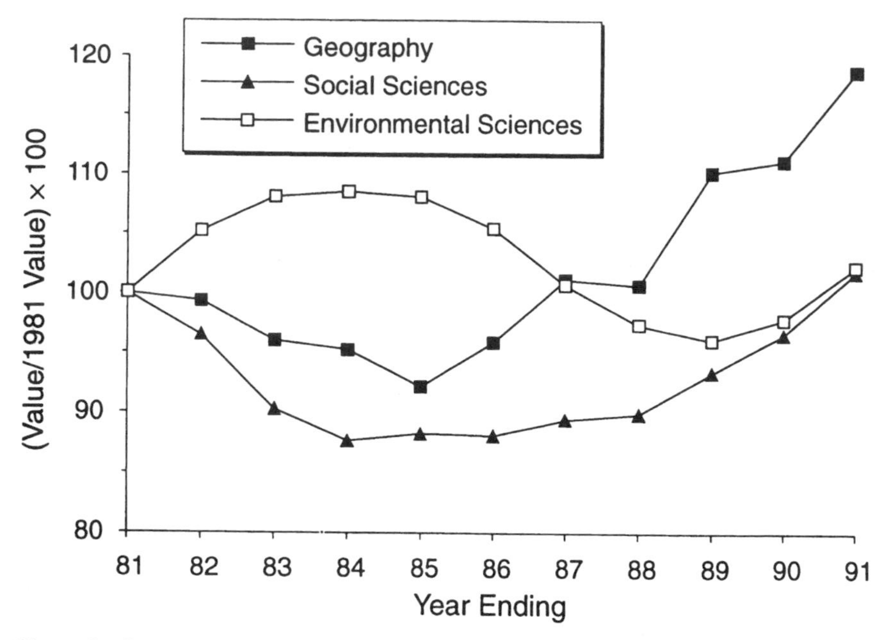

Figure 3: *Graduate enrollment in geography, social sciences, and environmental sciences, 1981–1991.*

we have graduate enrollment data, that added supply of graduate-level geographers probably is occurring already.

To obtain information about trends in undergraduate enrollments and, hence, the future supply of bachelor's-level geographers, we examined the number of "students in residence" in the *Guide to Programs of Geography in the United States and Canada* from 1986–1987 through 1993–1994. We eliminated those schools for which we were unable to obtain a complete record of enrollment data because we wanted to conduct a longitudinal investigation of how a constant set of schools performed through time rather than to obtain a series of cross-sectional studies of an ever-changing mix of programs. This elimination process reduced the number of schools to 174 from the 239 listed in the 1993–1994 *Guide*. The list of 174 includes 51% of all bachelor's-granting departments in 1993–1994, 88% of the master's departments, and 85% of the doctoral programs; thus, the experience of graduate departments will be overrepresented in the overall tallies. The average bachelor's-granting department contained 71 undergraduate students in residence, master's departments had 93, and doctoral departments had 104. If the missing cases display the same size patterns as the schools for which we have continuous data, they contain about 33% of the enrollment in undergraduate geography programs nationwide, and the 174 departments that form the basis of the study contain approximately two-thirds of geography's undergraduate enrollment.

The number of undergraduate students in residence in the 174 departments for which continuous data are available grew by 47% from 1986–1987 to 1993–1994. When these enrollment figures are normalized to their 1986–1987 levels and then viewed alongside the undergraduate degree figures discussed earlier, the two patterns are highly consistent (Fig. 4). Given that enrollments, as reported by the 174 schools in our study set, are a

Table 1 *Graduate Enrollments in Geography and Related Fields, Fall 1985 to Fall 1991*

Field	1985	1991	Percent Change 1985–1991
Geography	**2,836**	**3,785**	**33.4%**
Social Sciences	70,450	81,279	15.3
Anthropology	5,631	6,695	18.8
Economics	12,430	12,709	2.2
Sociology	6,567	8,314	26.6
Political Science	27,012	31,929	18.8
Environmental Sciences	15,591	14,747	–5.4
Atmospheric Sciences	964	968	.4
Geosciences	10,294	7,626	–25.9
Oceanography	2,081	2,293	10.2

Source: National Science Foundation, Division of Science Resources Studies. Academic Science/Engineering: Graduate Enrollment and Support, Fall 1991.

reasonably accurate guide to degrees granted in geography, it is logical to expect the accelerated enrollment in geography programs after 1992 to translate into higher degree production in 1992–1993 and 1993–1994.

Enrollment gains varied in a predictable fashion by region (Table 2). They were slower than the national average of 47% in the Northeast and Midwest, regions experiencing sluggish population growth and slow growth in student enrollments generally. Faster-than-average growth in enrollments was recorded in the South and West, areas of rapid population growth, net domestic in-migration, and immigration from abroad. The latter two regions accounted for almost two-thirds of the increase in geography enrollments nationwide.

Enrollment data also reveal that geography programs in Ph.D.-granting departments were the biggest generators of undergraduate enrollment increases. Between 1986–1987 and 1993–1994, enrollments increased by a stun-

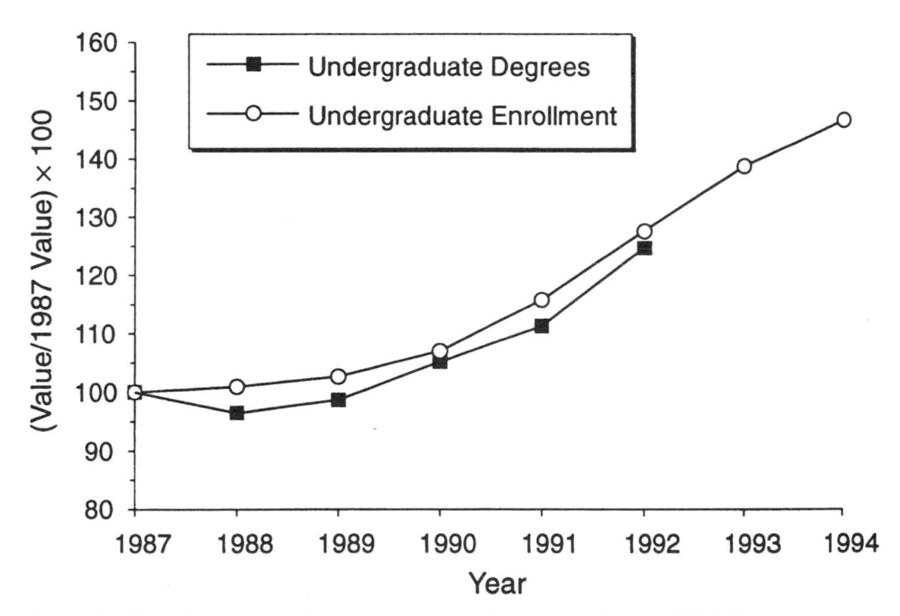

Figure 4: *Normalized undergraduate geography enrollments and degrees, 1987–1994.*

Table 2 *Undergraduate Enrollments by Region and Type of Geography Program*

	1986–1987	1993–1994	1986–1987 to 1993–1994
Total enrollment	10,743	15,752	46.6%
Region			
Northeast	1,839	2,453	33.4
Midwest	3,057	4,261	39.4
South	2,662	4,169	56.6
West	3,185	4,869	52.9
Type of program			
Undergraduate	2,224	3,339	50.1
Masters	5,196	7,095	36.5
Doctoral	3,323	5,318	60.0

Source: Authors' calculations based on analysis of AAG Guide to Programs of Geography in the United States and Canada *(1986–1987 to 1993–1994).*

ning 60% in Ph.D. departments, compared with 50% in undergraduate departments and 37% in master's programs. This finding probably reflects the growing emphasis on undergraduate education in major research universities nationwide.

The NCES has projected rates of growth in the number of degrees for higher education as a whole to the year 2002–2003 (NCES 1993b). Applying these rates to geography assumes that the discipline will reflect the trends prevalent in all of higher education. Our track record relative to these national norms, unfortunately, is not good. The number of degrees granted in geography declined throughout most of the 1970s and 1980s even though the total number of degrees granted by all fields grew. Moreover, since geography began its rebound in 1988, it has outpaced higher education as a whole in the number of degrees granted. NCES projects aggregates rather than individual disciplines or areas (social sciences, education, engineering, physical sciences, etc.) for precisely this reason. Some fields grow while others stagnate. Students "vote with their feet," moving easily and quickly among fields in response to changing labor market conditions, social values, and personal preferences (i.e., the recent experience of geology). Projections using 1985–1986 as a base year would have, for example, seriously overestimated degrees and enrollments in the physical sciences and underestimated those in the social sciences.

Nevertheless, we applied projections for higher education as a whole to geography (Table 3). We envision these projections as a benchmark against which we can assess the discipline's future performance rather than as hard-and-fast expectations. The numbers of undergraduate degrees in higher education are expected to rise until the mid-1990s, hold steady until 2000, and then increase again early in the next decade. Applying these projections to geography anticipates an increase in undergraduate degrees from 3,397 posted in 1990–1991 to 4,080 in 2002–2003.

The previous discussion of degree and enrollment trends in the field, however, suggests much faster growth in the number of geography undergraduate degrees. The figure for 1991–1992 was already above 3,800. Furthermore, if 13,701 "students in residence" in geography programs generated 3,808 degrees in 1991–1992, then the undergraduate enrollment figure of 15,752 in 1993–1994 should translate into 4,378 bachelor's degrees in that year, far more than higher education trends indicate for the entire projection period extending to 2002–2003.

Higher education projections are also problematic at the graduate level because more undergraduate degrees filter up through the system and generate more master's and Ph.D. degrees. Throughout the 1980s, the ratio of bachelor's to master's degrees hovered between 5 and 6 (Fig. 5). Assuming this ratio holds firm (and we have a long historical record to suggest that it will), we expect that the number of master's degrees will rise to 800 in 1993–1994, again a far larger figure than an extrapolation of higher education trends would indicate.

Because of their relatively small number, Ph.D. degrees vary more from one year to the next. Since 1980, the ratio of bachelor's to Ph.D. degrees has ranged from 22 to 29, averaging 25 (Fig. 5). Using this average ratio of

Table 3 *Projections of Bachelor's, Master's, and Doctoral Degrees in Higher Education and Geography, 1990–1991 to 2002–2003*

	Bachelor's Degrees		
	Higher Education		Geography
1990–1991	1,084,000		3397
	(+86,000)	7.93%	(+270)
1995–1996	1,170,000		3667
	(+15,000)	1.28%	(+46)
1999–2000	1,186,000		3714
	(+117,000)	9.86%	(+366)
2002–2003	1,303,000		4080
	Master's Degrees		
	Higher Education		Geography
1990–1991	337,000		622
	(+17,000)	5.05%	(+31)
1995–1996	354,000		653
	(–1,000)	–.28%	(–2)
1999–2000	353,000		651
	(+12,000)	3.40%	(+22)
2002–2003	365,000		673
	Doctoral Degrees		
	Higher Education		Geography
1990–1991	40,000		119
	(+1,100)	2.72%	(+3)
1995–1996	41,100		122
	(+500)	1.22%	(+1)
1999–2000	41,600		123
	(+200)	.48%	(+1)
2002–2003	41,800		124

Source: Gerald, D. E., and W. J. Hussar, 1993. Projections of Education Statistics to 2003. *National Center for Education Statistics.*

25 and an estimated 4,378 bachelor's degrees, we expect the number of Ph.D. degrees to be 175 in 1993–94.

To confirm the validity of these expectations, we asked departmental chairs, individuals who are at the front lines of enrollment management in colleges and universities, to assess past and future trends in numbers of undergraduate degrees. We heard back from 214 of the 376 chairs to whom we sent questionnaires, for a response rate of 57%. This overall response rate was composed of rates of 76% for institutions offering the Ph.D. in geography, 58% for schools offering master's degrees, and 49% for other institutions. Included in the latter category are some two-year institutions and schools without full-fledged geography departments. Many are part of a bureaucratic unit whose chair or director is probably not a geographer and, thus, far less likely to respond to highly specific questions about the training and occupational prospects of geography students. Our sample, therefore, underrepresents the views of the chairs whose programs account for a small proportion of the total number of geographers produced by institutions of higher education in the United States.

Chairs reaffirmed the sizable increases in degrees granted during the last five years (Table 4). Some 58% noted some increase in degrees over the last five years. Moreover, a majority of chairs were bullish about the prospects for future growth. When asked to assess the change in the number of geography degree graduates 6–10 years from now, 45% predicted increases between 10 and 25% and another 16% envisioned increases of more than 25%.

Departmental chairs also anticipate growth in the size of their graduate student bodies. Compared with the 1993–1994 academic year, chairs anticipated that graduate enrollments would climb 3% at the master's level and 9% at the doctoral level by 1998–1999.

Geographic Education Initiatives

To gain yet another perspective on the future supply of geographers, we asked several key Geographic Alliance Coordinators to com-

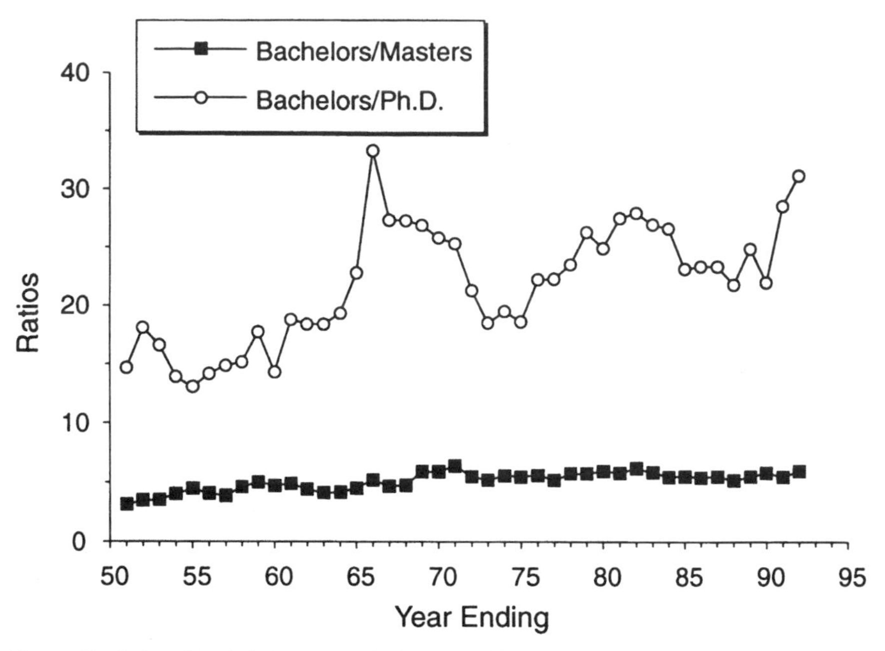

Figure 5: *Ratios of bachelor's to master's degrees and bachelor's to doctoral degrees in geography, 1951–1992.*

ment on the "trickle-up" effects of geographic education initiatives. To what extent, in other words, has the enhanced visibility of geography at the pre-collegiate level translated into more collegiate geography majors who will, in turn, constitute the future supply of new geographers? Interviews with A. David Hill from Colorado, Sidney R. Jumper from Tennessee, and Richard G. Boehm from Texas helped us to develop both a quantitative and a qualitative picture of these trickle-up effects. Keep in mind we are working from case studies of the success stories—states where Alliance efforts have been in place for some time, where geography is required for admission to the state's flagship universities (Colorado and Tennessee) or where it is part of the recommended high school program (Texas), and where the overall visibility of the discipline is high.

Tennessee provides us with a quantitative picture of what happened to high school geography enrollments in an environment that was favorable to geographic education. In 1982–1983, 5,535 students took world geography courses in Tennessee high schools. In 1986 the Tennessee Geographic Alliance was formed, and in 1989 the University of Tennessee added a "world geography or world history" admission requirement. By 1989–1990 world geography enrollments had soared to 18,487, and by 1992–1993, the last year for which we have data, they reached 28,733. Sidney Jumper, Geographic Alliance Coordinator and departmental chair at the University of Tennessee, reports that the increased presence of geography at the secondary level has affected the University of Tennessee's Department of Geography in a number of important ways. First, the department has gone from no incoming freshman majors each year to eight in 1992–1993 and four in 1993–1994. Second, undergraduate courses (service courses as well as those in the major) are oversubscribed. Third, the quality of students in the undergraduate program has improved, not so much as a result of better preparation, but from the fact that better students are now choosing geography as a field of study. And finally, the department

Table 4 *Chairs' Estimates of Past and Future Changes in the Number of Geography Degree Graduates*

	Change in Last 5 Years		Change 6–10 Years from Now	
Large increase (more than 25%)	37	(18.7%)	32	(15.8%)
Modest increase (10–25%)	77	(38.9%)	91	(45.0%)
Fairly stable (+/– 10%)	71	(35.9%)	41	(20.3%)
Modest decrease (10–25%)	9	(4.5%)	9	(4.5%)
Large decrease (more than 25%)	3	(1.5%)	1	(.5%)
Unable to say	1	(.5%)	28	(13.9%)
Total	198	(100.0%)	202	(100.0%)

Source: Departmental chairs' survey.

was the only unit in the University's College of Arts and Sciences to gain a new position in 1993–1994. Jumper attributes this in part to the department's ability to demonstrate community outreach in the form of geographic education activities.

The story is similar in Colorado. Although high school enrollments are not available there, we do know that 78 of 150 school districts in the state instituted geography classes in the past five years, many in response to the inclusion in 1988 of a geography requirement in the University of Colorado's admission standards. Effects at the university level were an increase in the number of incoming freshmen geography majors, more undergraduate majors overall (from 107 in 1988 to 175 in 1992), an avalanche of new students (1,000 per semester) in an introductory geography course used to fulfill the admissions deficiency in geography, and a new faculty line targeted to provide instruction in this introductory course.

Texas Geographic Alliance Coordinators also describe an increase in majors and in freshmen taking geography courses. At Southwest Texas State University (SWTSU) the number of incoming freshmen geography majors, a barometer of geography's presence in high schools, increased from zero in the fall of 1983 to 29 in the fall of 1993. The number of incoming freshmen taking geography courses also increased over the same period, from 30 to 250. According to Richard Boehm, chair of the SWTSU geography department, the preparedness of these new students has not changed appreciably. What has changed is their heightened awareness of geography as a field of study and as a potential career path. Trickle-up effects also extend to the graduate level at SWTSU, as Boehm reports recruiting ten geography teachers into the department's master's program.

These case studies suggest that college and university geography departments will feel the effects of geography's higher profile at the precollegiate level. From a labor market perspective, more enrollment leads to more majors (the vast majority of geography majors come indirectly to the discipline after taking and enjoying an introductory geography course), more majors mean more degrees, and more degrees generate an ever-larger pool of people with geography backgrounds seeking employment.

Before we extrapolate these forces to all states and all geography departments, however, it is important to remember that we picked several of the most active and successful Alliance departments. In addition, the widespread adoption of the Geography National Standards, the key to diffusing these success stories nationally, is anything but assured. In the most optimistic scenario, the standards will be widely adopted and implemented, and the Tennessee, Colorado, and Texas success stories will diffuse across the nation, stimulating continued growth of new majors and, eventually, new geographers entering the work force. In the less optimistic scenario, other states will be unable to repeat these experiences, adoption of the National Standards will be limited, and the number of future geographers will level off.

Future Supply Characteristics

Numbers alone do not tell the full story of future supply conditions in geography. What types of skills and interests do students now in the educational pipeline have? We asked departmental chairs about tracks or specializations in their undergraduate programs and the types of occupations for which their students were being prepared. Of the 212 geography chairs who responded to our survey, 139, or 65%, reported that their departments had at least one specialization. These specializations were led by programs in environmental/resource management, techniques (GIS, cartography, and remote sensing), and urban planning. These tracks appear to be designed to prepare students for the occupations in which geographers traditionally have found work rather than to develop their interests in regional geography or the systematic specialties like urban, economic, or physical geography that have traditionally formed the core of the academic discipline.

We also asked chairs to indicate the specific occupations for which students were being prepared (Table 5). The occupations indicated by the highest numbers of departments were (1) GIS/remote sensing specialist, (2) secondary school teacher, (3) cartographer, (4) environmental manager/technician, and (5) urban/regional planner. When chairs were asked to estimate the percentage of students in their programs who were preparing for particular occupations, a slightly different set of occupations emerges. Although GIS was by far the most popular occupational trajectory (involving almost 11% of students enrolled in programs offering GIS training), it was followed by environmental manager/technician and urban/regional planner. Although training as a secondary school teacher and cartographer, two of geography's longstanding career paths, were offered in a larger number of departments, these occupations involved a smaller share of currently enrolled students. In other words, whereas future teachers and cartographers are being trained in most departments, relatively few departments have high percentages of students preparing for these occupations.

Summary and Conclusions

Analysis of degree and enrollment trends in geography reveals the discipline's up cycle from the mid-1960s to the mid-1970s, its down period from the mid-1970s to around 1988, and its current rejuvenation. Data from a wide range of sources tracking a variety of indicators point consistently to significantly larger numbers of undergraduate majors and degrees and graduate majors and degrees during the early 1990s. To the extent that the number of geography majors and geography degree recipients correlates closely with labor force entrants, it is clear that the supply of new geographers has risen dramatically in recent years.

The evidence also supports continued growth in the supply of new geographers. The increasing numbers of students currently en-

Table 5 *Occupations for which Geography Students are Being Prepared (As Reported by Departmental Chairs)*

Occupation	Number of Programs Preparing Students for This Occupation	Median Percent of Program's Students
GIS/Remote Sensing Specialist	172	11%
Teacher-Secondary	157	6
Cartographer	156	7
Environmental Manager/Technical	153	8
Planner—Urban/Regional	152	8
Planner—Environmental	148	8
Earth Scientist	124	6
Teacher—Collegiate	122	4
Teacher—Primary	119	4
Climatologist/Meteorologist	118	2
Community Development Specialist	116	1
International Specialist	109	3
Planner—Transportation	106	2
Aerial Photo Interpreter	102	2
Ecologist	98	4
Land Economist/Real Estate Professional	98	1
Marketing Analyst	95	1
Administrator/Manager	88	4
Soil/Agricultural Scientist	86	1
Writer/Journalist/Editor	82	1
Other	33	12

rolled in undergraduate geography programs should translate into larger graduate enrollments and degrees granted. Widespread implementation of the Geography National Standards would mean more incoming geography majors, a heightened awareness of geography as a field of study, and generally more students taking geography courses. It appears that the discipline is doing a very successful job attracting new students who have come to see geography as a stepping stone to a satisfying and productive career. The obvious next question involves the extent to which these students are, in fact, finding work in the field. The second article in this series describes current labor market conditions for new geographers. ■

Literature Cited

Bureau of Labor Statistics. 1994. *Occupational Outlook Handbook*. Bulletin 2450. Washington, DC: U.S. Government Printing Office.

Goodchild, M. F., and D. G. Janelle. 1988. Specialization in the structure and organization of geography. *Annals of the Association of American Geographers* 78:1–28.

Hart, J. F. 1966. *Geographic Manpower: A Report on Manpower in American Geography*. Commission on College Geography Report 3. Washington, DC: Association of American Geographers.

Hart, J. F. 1972. *Manpower in Geography: An Updated Report*. Commission on College Geography Report 11. Washington, DC: Association of American Geographers.

Hausladen, G., and W. Wyckoff. 1985. Our discipline's demographic futures: Retirements, vacancies, and appointment priorities. *Professional Geographer* 37:339–43.

Janelle, D. G. 1992. The peopling of American geography. In *Geography's Inner World*, eds. R. F. Abler, M. G. Marcus, and J. M. Olson, 363–90. New Brunswick, NJ: Rutgers University Press.

Miyares, I. M., and M. S. McGlade. 1994. Specializations in "Jobs in Geography." *Professional Geographer* 46:170–77.

National Center for Educational Statistics. 1993a. *Digest of Education Statistics: 1993*. NCES 993-292. Washington, DC: U.S. Government Printing Office.

National Center for Educational Statistics. 1993b. *Projections of Education Statistics to 2003*. Washington, DC: U.S. Government Printing Office.

National Center for Education Statistics. 1994. *Digest of Education Statistics: 1994*. NCES 94-115. Washington, DC: U.S. Government Printing Office.

National Science Board. 1993. *Science and Engineering Indicators: 1993*. Washington, DC: U.S. Government Printing Office.

National Science Foundation. 1993. *Academic Science/Engineering: Graduate Enrollment and Support, Fall, 1991*. Washington, DC: U.S. Government Printing Office.

Suckling, P. W. 1994. National and regional trends in "Jobs in Geography." *Professional Geographer* 46:164–69.

Employment Trends in Geography, Part 2: Current Demand Conditions*

Patricia Gober
Arizona State University

Amy K. Glasmeier
Pennsylvania State University

James M. Goodman
National Geographic Society

David A. Plane
University of Arizona

Howard A. Stafford
University of Cincinnati

Joseph S. Wood
George Mason University

This paper, the second in a series dealing with employment trends in geography, focuses on current labor market conditions. Two windows on the current labor market are (1) the employment experiences of recent graduates of geography programs and (2) the activities of the Association of American Geographers Convention Placement Services (CoPS). The former provides a perspective primarily on the nonacademic labor market in geography and includes bachelor's, master's, and doctoral recipients of geography degrees. The latter covers both academic and nonacademic jobs but focuses on geographers who hold advanced degrees.

Recent Graduates

Very little is known about the interface between postsecondary education and the labor market in geography. There is not a definite point where people leave the university and enter the work force. Many students work full- or part-time while they attend the university, others continue their educations after receiving an undergraduate degree, and still others take time off before they begin their professional careers or work in occupations completely unrelated to their undergraduate degree program. Complicating matters further, in geography, direct paths into the labor force are not as evident as in disciplines like nursing, accounting, engineering, and other professional fields. The vast majority of geographers work in occupations that do not carry the title "geographer."

Recognizing how little we know about the labor market for individuals with bachelor's and master's in geography, we queried recent graduates of a random sample of geography programs about their labor market experiences. Are they finding work? What types of occupations are they engaged in and for what kinds of employers do they work? Are they making effective use of the skills obtained in their geography programs? Whenever possible, we linked the results of the survey to national data sets on educational outcomes in order to provide a contextual and comparative perspective for geography's experience.

We chose a random sample of U.S. geography programs from the *Guide to Programs of Geography in the United States and Canada* stratified by region (Northeast, Midwest, South, and West) and type of department (bachelor's, master's, and doctoral). One bachelor's, master's, and doctoral department was chosen from each of the four major regions for a total of 12 geography programs.[1] Departmental chairs provided us with lists of all graduates for calendar years 1990 through 1993. We mailed questionnaires to all 1,239 individuals on the lists and sent follow-up reminder postcards to those we had not heard from in two weeks. Our efforts yielded a total of 623 replies, 460 of them bachelor's recipients, 132 master's, 25 doctoral, and 6 unknown. Adjusting our population by the 38 questionnaires that could not be delivered by the U.S. Postal Service and for 16 respondents who turned out to be pre-1990 graduates, the overall response rate was a very respectable 53% (Table 1). The numbers in

*This article is excerpted from the appendix of the report of the National Research Council's Rediscovering Geography Committee.

Professional Geographer, 47(3) 1995, pages 329–336
Initial submission, November 1994; final acceptance, January 1995.
Published by Blackwell Publishers, 238 Main Street, Cambridge, MA 02142, and 108 Cowley Road, Oxford, OX4 1JF, UK.

Table 1 *Response Rates by Geography Program*

	Surveys Sent	Undeliverable/ Pre-1990 Graduates	Valid Responses	Response Rate
Dartmouth	77	0	44	57.1%
Towson State	146	2	70	48.5
Rutgers	85	8	50	64.1
IPUPI	33	3	16	53.3
Miami (Ohio)	63	4	30	50.1
Iowa	124	6	51	43.2
Jacksonville	36	2	14	41.2
SWTSU	392	11	200	52.5
George Mason	92	8	53	63.1
Sonoma State	54	4	29	58.0
Montana	44	2	27	64.3
Hawaii	93	4	43	47.7
Total	1239	54	623	52.6

the subsequent tables may not sum to 623 because not all of the respondents answered all of the questions.

We asked graduates about their current employment status and determined that 416 (68.7%) were working full-time, 37 (6.0%) were working part-time, 12 (2.0%) were in the military, 96 (15.6%) were going to school and working, 19 (3.1%) were going to school and not working, 19 (3.1%) were unemployed, and 9 (1.5%) were out of the labor force. Compared to national norms that apply to bachelor's recipients only, geography's recent bachelor's graduates were somewhat less likely to be engaged in full-time employment and more likely to be in school. We view this comparison with considerable caution, however, because the National Center for Educational Statistics' (NCES) employment categories are different from ours (NCES 1993). The NCES puts teaching and research assistants into a category called "enrolled in school" and considers them not to be in the labor force. This rule assumes that a student's primary activity is going to school. Moreover, it assumes that the jobs held by students are available only to those attending school and not to the labor force in general. Our categories allowed students (including graduate assistants) to specify whether they worked or not, thus acknowledging that individuals can be students and labor force members simultaneously. Although our categories are more complicated, they better reflect the increasingly fuzzy distinction between educational and employment status.

Among bachelor degree recipients, we discovered that more of the earlier than later graduates were employed full-time, fewer were employed part-time, and fewer were unemployed (Table 2). The trend in the percent in school is quite typical. There is an initial surge by those who go directly from a bachelor's to a master's program, then a falling off followed by an increase about two years out. This is a big decision point for many, and a commitment to return to school comes at about this time. The small numbers of unemployed may be deceiving because the hard-core unemployed are unlikely to respond to a survey such as ours.

Overall, these results suggest that geography graduates, like many of their counterparts in the arts and sciences, experience a period of transition and adjustment between the time they receive their undergraduate degrees and the point at which they settle into stable, full-time employment. This period is characterized by the pursuit of additional education, part-time employment, and internships. A typical career path for the new graduate in today's job market may involve temporary, contractual employment on a project-by-project basis during which the individual makes contacts with local planning agencies and consulting firms. Experiences gained through a series of these temporary or contractual arrangements often lead to full-time employment within several years. Although verification of this hypothesis awaits careful longitudinal analysis that follows a panel of recent graduates through their early work careers, our data are suggestive of such a pattern.

Differences in employment status by year of graduation probably do not result from varying labor market conditions. College graduates throughout the study period faced equally

Table 2 *Employment Status of Bachelor's Degree Recipients by Year of Graduation*

	Total (N=423)	1990 (N=85)	1991 (N=94)	1992 (N=115)	1993 (N=127)
Percent employed	76.1%	83.6%	76.6%	78.3%	69.3%
Full time	68.6	76.5	72.3	67.8	62.2
Part time	4.7	1.2	3.2	7.0	6.3
Military	2.8	5.9	1.1	3.5	.8
Percent in school	19.7	15.3	20.2	15.6	25.3
And working	15.4	12.9	14.9	13.9	18.9
And not working	4.3	2.4	5.3	1.7	7.1
Percent unemployed	3.1	1.3	1.1	4.3	3.9
Percent not in labor force	1.2	0	2.1	1.7	.8

Source: Recent graduates survey.

harsh employment prospects. If anything, conditions improved somewhat for the 1993 graduates. Nonetheless, this group was less settled than earlier cohorts into long-term careers. Nor is it likely that the recent increase in geography graduates described in the first article in this series has tightened the competition for jobs in the field to any measurable extent. The growth in degree holders is geographically widespread, and geographers represent but a small minority of applicants in any of the submarkets in which they compete.

To evaluate the occupational structure of geography degree recipients, we developed a list of 21 occupations with which we believed most geographers would identify and asked the respondents to indicate which occupation best applied to them. We also gave the option of an "other" category to accommodate those whose occupations were not well described by any of the items on our list. We judiciously recorded geographically oriented occupations from the "other" category into the categories into which we thought they best fit so that we would be left with an "other" category that contained only those occupations that are far removed from the field. For the most part, these jobs were professional in nature, including minister, social worker, insurance underwriter, financial analyst, lawyer, and pilot, but they also included less-skilled positions such as van driver, waiter, and secretary. For all degree recipients (bachelor's, master's, and doctoral), approximately one-third of those who were employed and who listed an occupation were engaged in jobs that are not closely identified with geography. The fact that two-thirds of the respondents were employed in fields directly related to geography is an encouraging figure in an era when many college graduates are working in occupations that do not even require a college degree and the vast majority of social science graduates are employed outside their areas of study.

The proportion of new geographers at the bachelor's level holding positions in closely related fields is high by social science standards. During the early 1990s, only 16% of social science bachelor's degree recipients nationwide were employed full-time in jobs closely related to their fields of study one year after graduation (NCES 1993). Among our 1993 graduates, this figure was 38%, and for many of them, a full year had not elapsed since graduation. Our estimate of geography's figure was higher than for arts and sciences disciplines as a whole (26%) but lower than the norm for professional and technical fields (48%) (NCES 1993).

Much of the preceding discussion isolated the experience of bachelor's degree recipients in order to draw connections to national data sets. The subsequent analysis grouped together all degree levels. Even with 623 respondents, we do not have a sufficient sample size to allow meaningful disaggregation by degree levels. At this point, we seek a general understanding of the types of jobs that geographers hold irrespective of their degrees.

Among those who were employed and listed occupations closely related to geography, respondents clustered into five predictable occupations: teacher (15.6%), environmental manager/technician (12.9%), GIS/remote sensing specialist (10.5%), cartographer (8.2%), and planner (6.7%) (Table 3). The categories correspond quite closely to the most common occupations for which current students are preparing (as reported in the survey of departmental chairs in Part 1 of this series), although GIS/remote sensing specialist ranked higher on the list of occupations for which current

Table 3 *Recent Graduates (All Levels) by Occupation*

	Number	Percent
Aerial photo interpreter	5	.9%
Administrator/manager	15	2.7
Cartographer	46	8.2
Community development specialist	2	.4
Demographer/population researcher	7	1.2
Earth scientist	14	2.5
Ecologist	8	1.4
Environmental manager/technician	73	12.9
GIS/remote sensing specialist	59	10.5
International specialist	5	.9
Land economist/real estate professional	6	1.1
Marketing analyst	11	2.0
Planner–environmental	16	2.8
Planner—urban/regional	14	2.5
Planner—transportation	8	1.4
Teacher—primary	17	3.0
Teacher—secondary	26	4.6
Teacher—collegiate	45	8.0
Writer/journalist/editor	3	.5
Other	184	32.6
Total	564	100.0

Source: Recent graduates survey.

students are preparing than on the list of where recent graduates are working. These five categories (or aggregates of categories in the case of teaching and planning) accounted for 54% of the respondents who listed their occupations and 80% of those who chose from the list of geographically oriented occupations.

In terms of employers, the breakdown was 34% in government (local, state, and federal), 40% in the private sector, 17% in educational institutions, and the remainder in nonprofit organizations (Methodist Church, NAACP, health and social services, environmental groups, etc.) and as self-employed consultants and small-business owners (Table 4). When respondents were disaggregated by how directly their work relates to geography, the private sector emerged as the dominant employer of those working at the margins and outside of the field. Still, some 45% of those respondents reported an occupation closely related to their geographic training. Graduates working in government and education were the most likely to be involved in geographic occupations.

We cross-classified type of employer by occupation to gain a clearer picture of kinds of work geographers do for different types of employer (Table 5). Cartographers were concentrated in the federal government and, to a lesser extent, in private industry. Environmental managers worked in state government, supervising the myriad new environmental regulations, and in the private sector for environmental consulting firms that have been established to meet these regulations. Some 44% of the GIS/remote sensing specialists worked in the private sector. Planners were strongly clustered in local and, to a lesser extent, in state government. Not surprisingly, teachers were employed in educational institutions. Our residual category of other geographic professions was more concentrated in the federal government, nonprofit organizations, and self-employed settings than the respondents as a whole.

The strong association between the "other" category and the private sector can be interpreted in three ways. The standard interpretation is that some geographers are unable to find work or are uninterested in working in the field after graduation and therefore seek out or settle for private-sector positions not traditionally associated with the discipline. A second perspective is that we have defined the notion of geographic occupations much too narrowly. Individuals who can use, display, and analyze geographic information to solve spatial and environmental problems can find jobs in all kinds of businesses, many with job titles that have not traditionally been associated with the field. Geography plays to the creativity, diversity, and flexibility of the contemporary labor market as much as it fills the standard niches of cartographer, planner, environmental manager, GIS/remote sensing specialist, and

Table 4 *Recent Graduates (All Levels) by Employer*

	Number	Percent of Total	Percent In Closely Related Fields
Local government	54	9.5%	79.6%
State government	60	10.6	85.0
Federal government	78	13.8	77.9
Private industry/business	227	40.0	44.9
Educational institution	96	16.9	90.6
Nonprofit organization	22	3.9	72.7
Consultant/self-employed	30	5.3	73.3
Total	567	100.0	67.4

Source: Recent graduates survey.

teacher. Many of our respondents saw themselves as performing geographic jobs even though this was not reflected in their job titles.

A third interpretation is that individuals working in fields other than those traditionally considered geographic find work in areas in which they have prior experience. Since our analysis neither holds constant the respondent's age nor classifies individuals on the basis of prior education or professional experience, we do not know the extent to which our survey is picking up prior preparation or work experience that leads to a nontraditional employment experience.

Greater confidence in our hypotheses about the recent labor market experiences of geographers requires more in-depth study of individual work experiences during the early- to mid-career years. Longitudinal analysis can verify the existence of a period of transition and adjustment when students move from part-time, temporary employment to full-time positions closely connected to their geographic training. Content analysis can decipher the potential relevance of the educational experience of students receiving degrees to tasks performed on the job. What if, for example, field-defined skills such as GIS, remote sensing, air photo interpretation, and cartography are not as important as the students' problem-solving and decision-making abilities? From a curricular standpoint, greater emphasis on professional practice, demeanor, and real world work experience may be more important than adding technical courses to academic curricula.

Convention Placement Service (CoPS)

A second glimpse at current labor market conditions in geography comes from the AAG's Convention Placement Service (CoPS). Held at the AAG's annual meeting, CoPS seeks to match job applicants with employers. Since 1989, the first year for which data are available, there has emerged a growing mismatch between the number of applicants and the number of interviews granted to these applicants (Fig. 1). In 1994, the 259 job seekers registered with CoPS attended only 156 interviews—an average of less than one interview per applicant.

Employers and job applicants who register with CoPS select from among a number of job

Table 5 *Graduates (All Levels) by Employer and Occupation*

Employer	Cartographer (N=46)	Environmental Manager/ Technician (N=73)	GIS/Remote Sensing Specialist (N=59)	Planner (N=38)	Teacher (N=88)	Other Geographic Professions (N=76)	Other (N=184)
Local government	8.7%	5.5%	13.6%	50.0%	3.4%	6.6%	6.1%
State government	8.7	27.4	10.2	23.7	3.4	11.8	4.9
Federal government	41.3	13.7	20.3	7.9	1.1	19.7	9.2
Private industry/ business	26.1	42.5	44.1	15.8	1.1	32.9	67.4
Educational institution	4.3	2.7	5.1	0	86.4	5.3	4.9
Nonprofit	2.2	1.4	1.7	0	3.4	13.2	3.3
Consultant/self-employed	8.7	6.8	5.1	2.6	1.3	10.5	4.3

Source: Recent graduates survey.

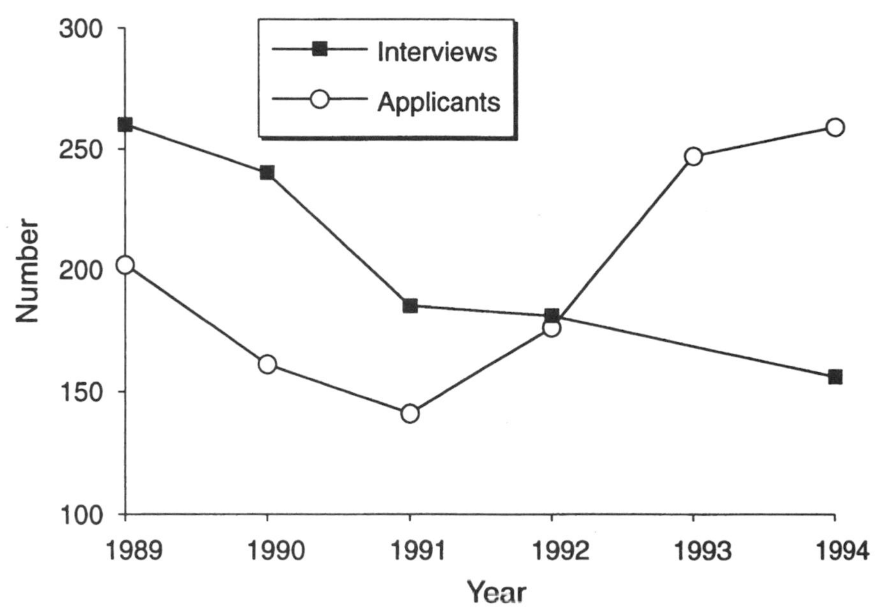

Figure 1: *Number of Applicants and Interviews Recorded by the AAG Convention Placement Service. Note: The data point for interviews in 1993 is not available. The box containing 1993 interview records was lost in the mail between Atlanta and the AAG office in Washington.*

categories. Many applicants and some employers list in more than one category. In 1994, 259 applicants accounted for some 556 listings, and 38 employers listed a total of 51 positions. Academic positions are included in two categories: an unfortunately large and heterogeneous physical/environmental/techniques group and a human/regional group. At the 1994 meeting, academic categories accounted for 43% of employment listings and one-third of the applicant listings (Table 6).

Applicants and employers for nonacademic jobs choose among four categories: physical/environmental (climatology, meteorology, geomorphology, hydrology, soils, biogeography, conservation, resources, ecology, environmental impact), planning (urban, transportation, land use, recreation, housing preservation, health, population), marketing/site selection/area studies (market research, real estate, public relations, advertising, regional expertise, travel consulting), and technical (cartography, remote sensing, photogrammetry, information systems, statistical and mathematical modeling). Applicants registered with CoPS were spread fairly evenly across the employment categories, but employers were concentrated in the nonacademic technical category—a sign of the strong link between GIS expertise and entry into the nonacademic labor market for geographers. There were a total of only six positions listed in the physical/environment, planning, and marketing/site selection/area studies classifications. While there were far more job applicants than positions even in the nonacademic technical category, the ratio of job seekers to positions was much lower than in the other nonacademic areas. For planning jobs, the ratio of applicants to jobs was more than 50 to 1!

Given the fact that these data apply only to individuals registered at the AAG annual meeting and employers attending this meeting, we make no claims that they are indicative of the

Table 6 *Applicants and Employers Registered with the AAG's Career Opportunities and Placement Service in March 1994*

	Applicant Listings	Employer Listings	Ratio Applicants/Employers
Academic			
Physical/environmental/techniques	85 (14.8%)	14 (27.5%)	6.1
Human/regional	111 (19.3%)	8 (15.7%)	13.8
Nonacademic			
Physical/environmental	98 (17.7%)	2 (3.9%)	49.0
Planning	107 (18.6%)	2 (3.9%)	53.5
Marketing/site selection/area studies	71 (12.3%)	2 (3.9%)	35.5
Technical	104 (18.1%)	23 (45.1%)	4.5

Source: Number of applicants and interviews recorded by the AAG Convention Placement Service.
Physical/environmental: climatology, meteorology, geomorphology, hydrology, soils, biogeography, conservation, resources, ecology, environmental impact, etc.
Planning: urban transportation, land use, recreation, housing, preservation, health, population, etc.
Marketing/site selection/area studies: market research, real estate, public relations, advertising, regional expertise, travel consulting, etc.
Technical: cartography, remote sensing, photogrammetry, information systems, statistical and mathematical modeling, etc.

total labor market for geographers. They do suggest a tighter labor market for new geographers at the master's and doctoral levels, and they signal the strong interest in techniques by potential employers of geographers.

Summary and Conclusions

The survey of recent graduates confirmed the continued importance of the five occupations in which geographers traditionally have been employed: cartographer, environmental manager/technician, GIS specialist, teacher, and urban/regional planner. These occupations accounted for slightly more than one-half of the respondents who reported their occupations. The strong association between "other" occupations and the private sector indicates that the private sector is the major destination of individuals who are not explicitly using their geographic training or are using it in occupations not traditionally associated with the field.

Results also revealed the quasi-professional position of geography relative to other social science fields. While only 16% of social science majors nationally were employed full-time in jobs closely related to their fields of study one year after graduation, this figure was 38% among the recent geography graduates in our survey. Geographers, although well placed by social science standards, fell below the norms of professional degree recipients, 48% of whom were employed in closely related fields.

Another finding involves the recent trend in the national labor market toward temporary, part-time employment as firms seek to reduce overhead costs and gain flexibility in managing the flow of work. Although our results need to be verified by careful longitudinal analysis, we picked up this trend in the experiences of recent geography graduates. Many newcomers to the labor market appear to face a period of transition between the time they receive their undergraduate degrees and the point at which they settle into stable, full-time employment. Geography programs can assist students in this transition period by encouraging or requiring internship experiences while in school, offering a senior seminar on the transition into the labor market, integrating more real-life experiences into the curriculum, and simply making students more aware of the challenges they will encounter in today's labor market.

Analysis of CoPS data suggested a tighter labor market for new geographers at the master's and doctoral levels and pointed to the strong link between GIS and other technical expertise and entry into the nonacademic labor market. Of the 29 nonacademic positions listed with CoPS in 1994, 23, or 79%, were in technical areas. Firms that participated in the 1994 national meetings appeared to be especially interested in graduate-level geographers with strong technical skills such as cartography, remote sensing, GIS, statistics, and mathematical modeling. Most geographers do not, however, learn these technical skills in an intellectual vacuum but in the context of the discipline's systematic specialties. The third in this series of articles explores the specializations and skills that will be demanded of geography graduates in the near future. ■

Note

[1]Because we do not know the total number of 1990–1993 graduates of geography programs by region or type of program, we do not know what proportion of all possible geography graduates our survey represents. Thus, our survey does not represent a probability sample in the strictest statistical sense.

Literature Cited

National Center for Education Statistics. 1993. *Digest of Education Statistics: 1993.* NCES 993-292. Washington, DC: U.S. Government Printing Office.

Employment Trends in Geography, Part 3: Future Demand Conditions

Patricia Gober
Arizona State University

Amy K. Glasmeier
Pennsylvania State University

James M. Goodman
National Geographic Society

David A. Plane
University of Arizona

Howard A. Stafford
University of Cincinnati

Joseph S. Wood
George Mason University

The third and final article in this series about employment conditions in geography addresses the issue of future demand in both academic and nonacademic settings. To gain an understanding of future demand conditions in colleges and universities, we projected the retirement of AAG members by topical specialty and then matched these retirement trends with a profile of new faculty searches as reported by geography department chairs. We assessed the likely future demand for geography teachers at the precollegiate level through a survey of Geography Alliance Coordinators about teacher certification requirements and the education environments in their respective states. We speculated on how the kinds of jobs geographers do will be affected by changes now underway in the national and global economies. And finally, we conducted a small telephone survey of AAG corporate sponsors to determine how future business trends will affect the demand for geographers.

Academic Job Market

Retirements

The majority of new academic jobs in geography will involve replacements on existing lines rather than new lines. Of the 340 "likely" searches during the next five years reported by the departmental chairs who responded to our survey (See Part 1 in this series for a more complete discussion of this survey), 240 (72%) will involve replacement lines. Only 95 (28%) will entail new positions. Replacements occur when faculty members change jobs, leave the academy, retire, and die. In this section our goal is to focus on retirement—to estimate how many geographers will retire in the next five to ten years, to determine their likely specialties, and to access how the discipline will change as a result of their retirement.

One of the universal laws of demography, not to mention nature, is that every individual grows older one year at a time. This inexorable

*This article is excerpted from the appendix of the report of the National Research Council's Rediscovering Geography Committee.

Professional Geographer, 47(3) 1995, pages 336–346
Initial submission, November 1994; final acceptance, January 1995.
Published by Blackwell Publishers, 238 Main Street, Cambridge, MA 02142, and 108 Cowley Road, Oxford, OX4 1JF, UK.

process allows demographers to estimate with a high degree of accuracy how many people will be born, die, migrate, marry, and divorce each year. Processes that are sensitive to age, like retirement, can be accurately predicted by examining the size and characteristics of the cohort entering ages when retirement is likely. We applied this logic to AAG membership data and focused on geographers 50 years of age and older.

As of January 1, 1994, the cohort older than 50 years was predominantly male. Only 13% of this group was female compared with 27% of all AAG members and 33% of those under 50 years. Another perspective is afforded by examining AAG topical specialties by age of members. The AAG's 51 proficiency categories were collapsed to 25 groups. Members were permitted to indicate up to three topical proficiencies; some indicate three, some two, some one, and some did not record any specialty. We used only the first proficiency indicated by each member in our tabulations. If there is a systematic bias toward members first reporting their main topical specialty, then this counting procedure captures this bias. If there is no bias, this simplification provides an accurate sample of specialties of AAG members. Multiple counting of topical specializations would have overstated the number of persons in each category expected to retire.

The most popular specialties overall (with more than 200 members claiming them first) were cultural/historical, economic, cartography/photogrammetry, conservation/land use, GIS, applied/planning, climatology/meteorology, physical, urban, and geomorphology (Table 1). In comparing the number of individuals born before 1940 (those likely to retire in the next ten years) with those born after 1959 (likely to replace retiring members), the biggest discrepancy occurs in the GIS specialization. Only 18 individuals with a GIS specialization were born before 1940, while 182 geographers born after 1959 claim this as their first topical specialty. In addition, more younger than older geographers associate with cartography/photogrammetry, conservation/land use, and the more specialized subfields of physical geography such as biogeography, climatology, and geomorphology. More older than younger geographers claim specializations in cultural/historical geography, agricultural/rural geography, regional geography, geographic thought, and geographic education.

Table 1 *AAG Membership by Year of Birth and Topical Specialty*

	Birth Year					
	After 1959	1950–1959	1940–1949	Before 1940	Total	After 1959–Before 1940
Agricultural/Rural	26	48	37	45	156	–19
Applied/Planning	64	75	101	62	302	2
Biogeography	62	78	38	18	196	44
Cartography/Photogrammetry	166	151	99	81	497	85
Remote Sensing	40	45	24	18	127	22
Climatology	82	93	61	45	281	37
Conservation/Land Use	151	127	110	91	477	60
Cultural Ecology	42	34	30	18	124	24
Cultural/Historical	135	124	190	182	631	–47
Economic	152	201	159	118	630	34
Geographic Education	27	29	30	33	119	–6
GIS	182	118	67	18	385	164
Geographic Thought	2	6	1	14	23	–12
Geomorphology	58	86	50	30	224	28
Physical	66	72	75	54	267	12
Political	40	42	38	36	156	4
Population	17	30	39	23	109	–6
Quantitative	8	8	15	6	37	2
Recreation/Tourism	10	12	12	13	47	–3
Social	34	35	20	12	101	22
Transportation/Communications	14	16	22	11	63	3
Urban	53	57	71	51	232	2
Water Resources	19	30	14	14	77	5
Other	20	18	37	39	114	–19
Regional	6	10	11	17	44	–11

Source: 1993 AAG membership data.

It is increasingly difficult to predict retirement rates for college and university faculty members or for any occupation because Congress, in amending the Age Discrimination in Employment Act in 1986, abolished the concept of a mandatory retirement age. Higher education was given special treatment in the sense that the act's protection did not apply to faculty over 70 years of age, but this provision was phased out in 1993 (Swan 1992). Today college and university faculty members have a great deal of latitude in the decision about when and under what conditions to retire. We used recent patterns of retirement in the AAG to predict the number of members who would retire in the near future. Retirement rates for 1993 AAG members were calculated by dividing the number of members reporting themselves as retired by the total number of members in each age class. These rates were then used to estimate the number of new retirees (a) during the period 1994–1998 and (b) for the period 1994–2003.

The percentages already retired were calculated using the 1993 data for the entire membership, by five- and ten-year age groups. The difference in retirement percentage between one age group and the next younger age group was multiplied by the number in the younger age group. This produces the number of new employment exits as each group progresses through the next five or ten years of their work careers. The 1993 retirement rates for the entire membership were applied to each of the subgroups. We assumed, therefore, that members of gender and specialty subgroups will retire at rates comparable to the AAG as a whole.

We estimate that 256 AAG members working at the start of 1994 will retire in the next five years and 617 in the next ten years. Most of the retirees will be male: 225 (88% of the total) in the next five years and 535 (87%) in the next ten years.

Not surprisingly, the cultural/historical category will account for more retirees than any other category (Table 2). Other specialties with many potential retirees are economic, conservation/land use, cartography/photogrammetry, applied/planning, physical, urban, climatology, and agricultural/rural.

New Faculty Hiring

Geography departments may or may not choose to replace retirees with individuals who have similar expertise. To gain an understanding of the expected future hiring practices of geography departments, we asked departmental chairs to indicate the specialty of new faculty hires. We used the same categories as

Table 2 *Retirement Projections by Topical Specialty*

	Next 5 Years	Next 10 Years
Agricultural/Rural	9	21
Applied/Planning	14	32
Biogeography	4	10
Cartography/Photogrammetry	15	37
Remote Sensing	3	10
Climatology	8	22
Conservation/Land Use	18	42
Cultural Ecology	4	9
Cultural/Historical	36	86
Economic	24	50
Geographic Education	8	18
GIS	3	13
Geographic Thought	3	6
Geomorphology	6	16
Physical	11	27
Political	8	16
Population	5	12
Quantitative	2	4
Recreation/Tourism	2	8
Social	2	6
Transportation/Communications	2	6
Urban	8	23
Water Resources	2	7
Other	8	20
Regional	3	6

Source: Projections based on 1993 AAG membership data.

***Table 3** Demand for Faculty Specialty Areas as Reported by Departmental Chairs and Specialties of Retiring AAG Members*

Specialty	Number of 5-Year Retirees (a)	Percent of Total (b)	Number of Likely Searches (c)	Percent of Total (d)	Ratio (d)/(b)
Agricultural/Rural	9	4.4	3	.5	.1
Applied/Planning	14	6.8	25	4.8	.7
Biogeography	4	2.0	13	2.5	1.3
Cartography/Photogrammetry	15	7.3	37	7.1	1.0
Climatology/Meteorology	8	3.9	28	5.3	1.4
Conservation/Land Use	18	8.8	51	9.8	1.1
Cultural Ecology	4	2.0	6	1.1	.6
Cultural/Historical	36	17.6	26	5.0	.3
Economic	24	11.7	36	6.9	.6
Geographic Education	8	3.9	13	2.5	.6
GIS	3	1.5	89	17.0	11.7
Geographic Thought	3	1.5	5	1.0	.7
Geomorphology	6	2.9	12	2.3	.8
Physical	11	5.4	41	7.9	1.5
Political	8	3.9	11	2.1	.5
Population	5	2.4	9	1.7	.7
Quantitative	2	1.0	17	3.3	3.3
Recreation/Tourism	2	1.0	11	2.1	2.2
Remote Sensing	3	1.5	30	5.7	3.9
Social	2	1.0	7	1.3	1.4
Transportation/Communications	2	1.0	7	1.3	1.4
Urban	8	3.9	19	3.6	.9
Water Resources	2	1.0	19	3.6	3.7
Other	8	3.9	7	1.3	.3
Total	205	100.0	522	100.0	

for the retirement study so that we could compare the specializations of likely retirees with those of likely searches (Table 3). Unlike the retirement analysis of AAG members, department chairs were allowed to indicate more than one desired specialty area for "likely" searches (those with an estimated probability of approval of greater than 50%).

Of all the likely searches, the most common specialty desired will be GIS (17%), followed by conservation/land use/resource management/environmental/hazards (10%), physical (8%), cartography/photogrammetry (7%), and economic/economic development/regional development/location theory/medical (7%) (Table 3). The cultural/historical category, which has the most retirees, is expected to be advertised for in only 5% of the likely searches in the next five years. The intended GIS hires will reflect a significant shift in emphasis area for many departments since so few of the likely AAG-member retirees list it as their primary field of interest.

The ratio of the percentage of likely searches in a given specialty to the percentage of retirees in that specialty is a rough indicator of the extent to which that specialty will increase or decrease its representation in geography departments as a result of faculty turnover (Table 3 and Fig. 1). A large ratio shows that a specialty is more highly demanded in the current market for new faculty than it is represented in the upcoming batch of new retirees. The highest ratios demonstrate the strength of technical specialties like GIS (11.7), remote sensing (3.9), and quantitative (3.7) and of physical geography (1.5) and its subspecialties of water resources (3.6), climatology/meteorology (1.4), and biogeography (1.3). The lowest ratios are in agricultural/rural (.1), cultural/historical (.3), political (.5), economic (.6), cultural ecology (.6), and geographic education (.6). While the hiring practices of collegiate geography departments appear to be highly responsive to the changing technical nature of the discipline, chairs of those departments indicated relatively little interest in hiring new faculty with explicit expertise in geographic education.

Precollegiate Geography Teachers

To gauge the future demand for geography teachers at the precollegiate level, we asked Geography Alliance Coordinators to respond to a series of questions about teacher certification requirements and the education environment of their states. We received replies from

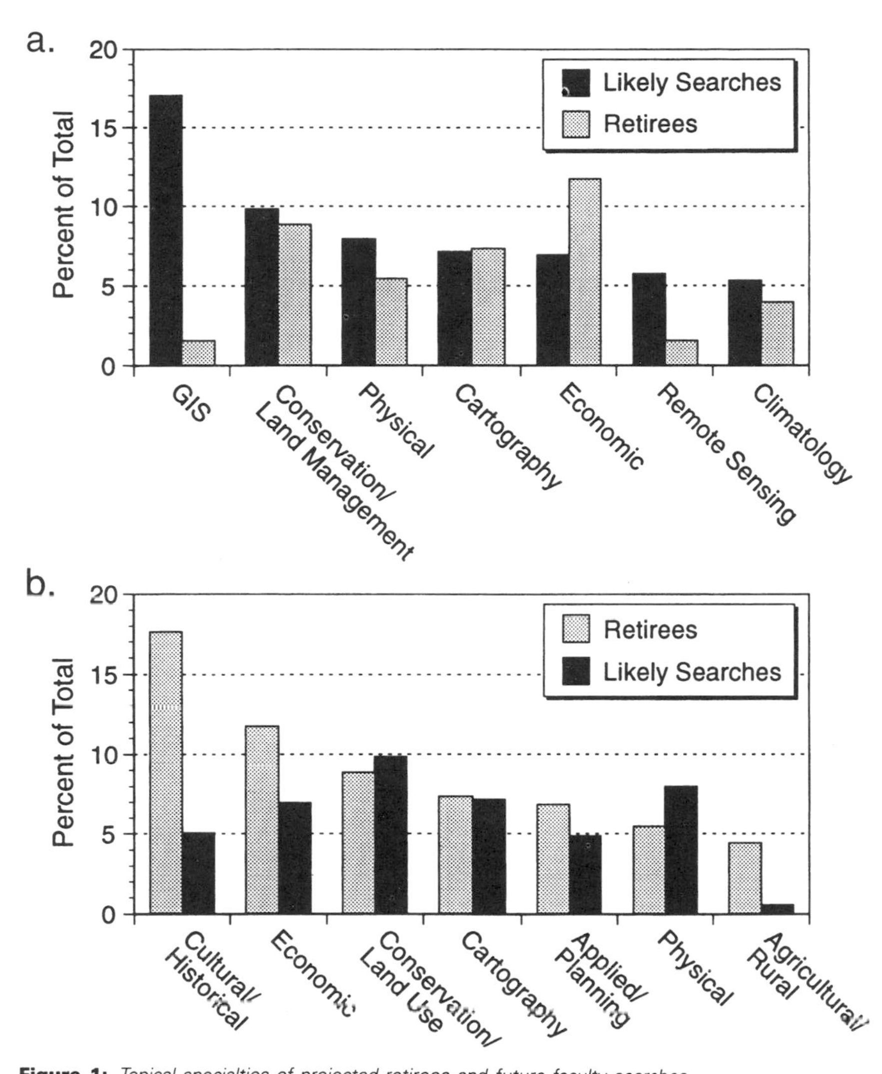

Figure 1: *Topical specialties of projected retirees and future faculty searches.*

coordinators in 37 of 50 states and the District of Columbia for a response rate of 73%. The crux of our questions involved state-level certification requirements. The extent to which geography courses are required and the rigor with which these requirements are enforced will influence the number and character of geography teachers demanded in the future.

What we discovered first is that education remains, as it was intended by Thomas Jefferson, a local matter. We received, quite literally, 37 different responses to our questions about certification and the local education environment. The Alliance Coordinators told us, in essence, that the future demand for geography teachers depends upon whether local school districts require geography courses, indeed, whether they offer any geography courses at all; whether the state mandates a geography requirement; whether universities require geography for admission (the Universities of Colorado and Tennessee as well as the Minnesota State University System do); whether there is a separate certification for geography or whether geography falls only under the umbrella of the social sciences; whether individual universities impose geography requirements over and above state standards; how individual universities choose to apply state requirements; and how many (if any) geography courses are required for certification. Should it occur, the full implementation of the National Geography Standards will be played out across a highly disparate landscape of geographic education, and it is onto that landscape that we must gauge the future demand for geography teachers.

Acknowledging the different settings in which the Alliance coordinators work, we did discover some common ground with respect to the future demand for geography teachers. We asked coordinators to "evaluate the present education environment in your state with respect to the future employment of geography teachers" over the next three and ten years. Responses reflected a cautious optimism about future demand (Table 4). Two-thirds of the coordinators anticipated some increase in the demand for geography teachers within the three-year horizon and 72% within the ten-year time frame. For the latter period, 35 of the 36 respondents who answered this question anticipated some increase, either modest or large.

But when coordinators were asked to express their expectations about the future demand for geography teachers in an open-ended question about the "key factors" affecting the future demand for geography teachers, a more qualified picture emerges. With a few notable exceptions, the coordinators do not expect large increases in the number of trained geographers teaching geography classes at the precollegiate level. Three interrelated themes underlie their qualifications. First, many argue that we should not view geography as a free-standing subject but as part of an integrated social studies curriculum. The demand for geography teachers will not change significantly. What will change is the idea that geography is part of the core of the training of every elementary teacher and secondary social studies teacher.

In one way or another, many of the coordinators also acknowledged the realities of the certification structure (that often little geography course work is required to teach geography) and the lack of connection between someone who teaches geography in the schools and someone who holds a degree in the field. In the short term, increased demand for geography in the schools will result in the transfer of teachers from other disciplines or grade levels; it will not result in large increases in the demand for individuals who hold degrees in geography or even have significant course work in the discipline. In the long run, if the National Geography Standards are widely accepted, there will be sustained improvement in the preparation of people teaching geography, and the separation between teachers of geography and geography degree holders will gradually disappear. We picked up the difference in the coordinators' long- and short-term perspectives when greater optimism was expressed at the ten than at the three-year time horizon.

And finally, some coordinators mentioned the age structure of teachers and the institutional framework of local school districts as

Table 4 *Geography Alliance Coordinators' Assessment of the Future Employment Prospects for Geography Teachers*

Response	Next 3 Years	Next 10 Years
Large increases	2 (5.6%)	9 (25.0%)
Modest increases	24 (66.7%)	26 (72.2%)
No change	9 (25.0%)	1 (2.8%)
Modest reductions	1 (2.8%)	1 (0%)
Large reductions	0 (0%)	0 (0%)

Source: Summary of Geography Alliance Coordinators' responses to the question: "How would you evaluate the present education environment in your state with respect to the future employment of geography teachers? Over the next three years? Over the next ten years?"

factors affecting the future demand for geography teachers. In one district in Minnesota, three-quarters of the secondary social studies teachers are in their 50s and are looking forward to early retirement. In other areas, the demographics do not allow for rapid turnover of the work force. A related concern among many coordinators is that institutional inertia will not permit rapid growth. Their perception is that social studies education traditionally has been dominated by historians who would not readily cede control of personnel and curriculum to geographers.

On the basis of this survey, we do not expect that the demand for geography degree holders will increase significantly in the short and midterm, as a result of geographic education initiatives. Geographic Alliance activities, along with national-level geographic education initiatives, will almost certainly result in a higher visibility for geography within the curriculum and in more geography courses being offered, but most of this material and most of the courses will be taught, at least in the short run, by transfers from other fields. The full and widespread implementation of the National Geography Standards, if it occurs, would have a deeper and more long-lasting effect on the long-term labor market for geography teachers.

Bureau of Labor Statistics Projections

The Bureau of Labor Statistics (BLS) forecasts employment change to the year 2005 by occupational and industrial classifications. Unfortunately, geographers do not fit easily into the established classification systems. BLS separates teachers into elementary, secondary, and college and university faculty but does not distinguish among academic disciplines or even between sciences and social sciences. The category of urban and regional planners captures geographers who are urban and regional planners and transportation planners, but environmental planners are grouped, for statistical purposes, with architects. GIS/remote sensing specialists are put into a large group of systems analysts and computer scientists. Cartographers are classified with surveyors but comprise a small proportion of the total category. Environmental managers/technicians are poorly defined by the very heterogeneous categories of "other physical scientists" (for managers) and "other engineering technicians" (for technicians).

Nevertheless, we looked at the BLS data to gain a broad understanding of future employment trends in those job classifications in which geographers are clustered (Table 5). BLS forecasts total employment growth between 1992 and 2005 using low, moderate, and high employment growth scenarios. For the national economy as a whole, moderate projections call for all occupations to grow by 22%. Professional and technical jobs involving high levels of skill and education are expected to grow faster than average. There is an enormous range in projected growth in the job classes frequented by geographers. That the class of systems analysts and computer scientists will experience the fastest growth fits with all the other evidence that we have collected pointing to GIS as a high-growth specialty within geography. In contrast, the class of surveyors, which contains cartographers, is expected to grow slowly, although it is questionable whether geographer/cartographers, who are much more closely related to GIS graduates than to surveyors, are truly represented in this category.

Projected employment growth is higher than average in the "other physical science" category in which our environmental managers are classified. Environmental legislation such as the Clean Air Act, Clean Water Act, Superfund Act, Resource Conservation and Recovery Act, and National Environmental Policy Act spawned new environmental regulations. These regulations, in turn, promulgated a new industry of environmental regulators working at various levels of government and environmental consultants helping the private sector to meet the regulations. At the same time, the number of urban and regional planners and architects is expected to grow at about the national average.

Higher than average growth is predicted for secondary teachers as a whole, and geography's favorable position relative to national education trends implies that this may be another growth area for new geographers. We need, however, to temper any optimism in this regard with all the caveats raised by Geographic Alliance Coordinators in assessing the future

Table 5 *Employment Projections to 2005 by Occupation*

	1992 Number ('000)	1992–2005 Employment Change Percent		
		Low	Moderate	High
All Occupations	121,099	15	22	28
Professionals	16,592	30	37	43
Architects	96	19	25	41
Surveyors	99	6	13	28
Computer systems analysts	455	96	110	120
Urban and regional planner	28	16	23	29
Other physical scientists	30	37	46	54
Elementary teachers	1,456	16	21	26
Secondary teachers	1,263	30	37	42
College and university faculty	812	20	26	31
Technicians	4,282	24	32	38
Other engineering technicians	392	7	16	25

Source: Silvestri, G. T. 1994. Occupational Employment: Wide Variations in Growth. The American Work Force: 1992–2005. *Bureau of Labor Statistics, Bulletin 2452.*

demand for geography teachers at the secondary level.

The demand for collegiate-level faculty and elementary teachers is expected to grow at rates about average for the economy as a whole. Geography's ability to make inroads at the collegiate level depends upon the discipline's ability to compete with related fields for a share of very limited budgets. The majority of vacant positions will result from replacements on existing lines rather than from new lines.

The BLS occupational projections for 1992–2005 are based on extrapolations of past labor market trends. The period immediately preceding these projections was characterized by a national recession with productivity levels down, part-time employment growing, savings levels persistently low, and medical and health care expenditures rising as a share of the gross domestic product (GDP). Since these occupational projections were announced, a number of these trends have changed (Kutscher 1993). The national economy is no longer in recession, although job growth is slow by historic standards. Productivity is on the rise, but this is partly due to the fact that firms are reducing labor forces while maintaining the same level of output. Part-time employment continues to grow and is becoming prevalent in sectors other than retail services. The share of GDP spent on health care appears to have peaked, yet the current debate in Congress renders moot almost any projection on longer-term impacts. Given the volatility of the current situation, coupled with the poor fit of geographers to BLS job categories, the BLS projections provide only the most general barometer of future labor market conditions for geographers.

National Economic Trends

We speculate on the way geographers will be affected by a number of important changes underway in the national economy. The trend toward increasing computerization continues to hold. Entry-level jobs assume computer literacy beyond simple word processing and data management. At the same time, basic input activities are increasingly being automated with the help of scanning technology and other innovations. Employers want employees to be able to use data for analysis, not just for information compilation (Farnham 1993).

Geography graduates will also be influenced by the growing trend toward part-time employment and contingent labor contracts (Fierman 1994). We noted earlier the tendency for recent graduates to be employed part-time immediately after receiving the bachelor's degree. Departmental chairs further verified the growing connection between geography graduates and part-time employment. When queried about the future hiring of part-time faculty, 37% thought the demand for such faculty in their departments would increase during the next five years. Only 13% predicted that they would be less reliant on temporary or part-time faculty.

Part-time employment provides flexibility and reduces overhead costs for firms and organizations. While once primarily the purview of firms in the retail sector, part-time work is increasingly found in sectors other than services. Corporations are downsizing and subcontracting activities formerly performed inside the corporation. Downsizing and subcontracting are occurring at all levels, including blue-collar, administrative, and executive-level jobs. Thus, a person's first job may not be permanent or even full-time. Moreover, it is less likely to require skills specific to the firm. By implication, firms are looking for people with strong general backgrounds, a range of skills, and the ability to work efficiently in many different settings.

Another important trend with the potential to affect geographers is the growing tendency for private-sector firms to use pools of labor found around the world to undertake such service activities as software design and computer programming (Martin 1992). For example, a corporation may hire skilled workers in low-cost countries such as India and Ireland to perform tasks that have heretofore been conducted on shore. Companies currently hiring graduates with GIS training may be able to find cheaper, equally skilled workers in other English-speaking countries. Lest we think this trend cannot have an impact on our field, keep in mind that major software companies use English-speaking labor forces around the world to answer user problems and to provide technical assistance. Companies like Texas Instruments and International Business Machines and construction companies such as Bechtel have established white-collar technical support staff in developing countries to take advantage of skilled labor at low cost.

The labor market for geographers may be affected by the continued decline in the defense sector, which is thrusting thousands of engineers and technicians into a very weak aerospace labor market. Consequently, these highly skilled, computer-literate individuals with significant problem-solving capabilities may compete for jobs formerly held by geographers with computer skills. Much like the 1970s, when the end of the Vietnam War put thousands of technical workers out of jobs, displaced defense workers have filtered into the public sector, running data processing and other technical operations. The extent to which these workers compete for geographers' jobs depends on their ability to gain necessary skills. Given significant federal government funding for the retraining of aerospace workers (on the order of $100 million), there is a strong likelihood of a growing supply of technical job seekers, some of whom will get jobs that geographers might have filled.

Also relevant to geography are privatization trends in government employment. Government at all levels is contracting for services, particularly services that require high-cost, in-house staff. The flip side of contracting is the growth in private-sector firms providing technical services. Data are too aggregate to discern whether consulting firms use contingent workers on government contracts; however, given that government contracts are in all likelihood small in size and short-term in nature, part-time work may be one means by which contract firms manage the flow of jobs.

What do these trends mean for the future employment of geographers? Geographers, like other highly skilled workers, increasingly are competing in highly volatile and uncertain labor markets. Moreover, the cohort of new geographers who will enter the labor market in the next decade will be forced, by their sheer size and by privatization trends in the public sector, to seek out positions in the private sector not previously associated with or held by geographers. Success in the current labor market seems to be tied to computer skills, a solid generalist background, and the ability to adapt to a wide range of job tasks and employment situations.

AAG Sponsors

We conducted a small survey of employers of geographers through structured interviews with AAG sponsors. These sponsors include book publishers, map and atlas publishers, geographic applications software developers, and geographic research applications specialists. In a written letter we invited representatives of the AAG's 12 corporate sponsors to communicate their views about the future demand for geographers and, more specifically, about the types of expertise that will be required for geographers to keep up with the ever-changing demands of their businesses. We were successful in conducting telephone interviews with eight of the 12 corporate sponsors. Each respondent was asked the following questions:

1. In what types of work, if any, do you employ geographers?
2. What are your impressions of the strengths and weaknesses of the training these individuals receive?
3. What are the major trends in your business that will affect the demand for geographers, and what forces generate these trends?
4. How can geographic training be improved to accommodate demands generated by these trends?

Not all corporate sponsors hire geographers. Yet each has a vested interest in the vitality of the discipline and was optimistic about the future demand for geographers. Corporate sponsors employ geographers in three ways broadly characterized as applications, customer service, and training. Applications geographers, the largest number, may be technicians, analysts, or managers. They are hired to provide geographic information in the form of vast data sets and maps and often to analyze this information using imagery analysis, GIS, cartography, photogrammetry, and spatial modeling. Corporate sponsors also hire geographers to train others and provide customer services for addressing such applications questions.

Sponsors prefer candidates who can act as decision makers, not just technicians. Knowing how to employ data is considered more important than knowing the technical specifications of data sources, software, or hardware. Sponsors repeatedly told us that they could hire engineers for such technical expertise. Geographers are valued for their mix of technical skills with a broad-based, multidisciplinary background. Geographers bring expertise in environmental and spatial problem-solving, and they are adaptable to a wide range of employment situations.

In the view of corporate sponsors, the primary trends affecting the future demand for geographers are the continued and widespread adoption of GIS technologies and the explosive growth in global positioning systems (GPS). Both have a variety of applications, including transportation design and development and utility planning, suggesting the need for geographers to take more courses in civil engineering. Other sponsors noted the need for geographers to merge GIS technology with visualization technologies.

While recognizing the pivotal role of GIS in driving the future demand for geographers, the sponsors expressed concern that GIS will lose its geographical identity—that applications will become less geographical and more like information processing. They also worry that increased technological capacity, such as hyperspectral and high-resolution spatial imagery, will bog down systems. Geographers able to understand appropriate uses, not simply more sophisticated uses, will be in demand.

Two themes emerged in response to our question about the weaknesses in the training of geographers. Some sponsors felt they were getting too many "button pushers" who know "cookbook" applications but are unable to work through a problem from start to finish. A related theme involves poor writing skills. It is clear that geographers' niche in the labor market comes from the ability to combine technical expertise with a broad training in the liberal arts stressing decision making, communication, and critical thinking.

We make no claims that the views of eight sponsors are indicative of the private sector as a whole. Instead, we offer this information as one more piece in the very complicated puzzle of future labor market conditions in the field. We note the consistency in the theme of GIS as the handle with which new geographers gain access to the labor market.

Summary and Conclusions

This scan of future demand conditions questions whether there will be sufficient jobs for the explosion of new geography majors currently in the educational pipeline (see Part 1 in this series). Positions at the collegiate level will be primarily replacement lines. In the short turn, it is unlikely that large numbers of new geographers will be needed at the secondary level, although this situation may change depending upon the success and comprehensiveness of the new National Geography Standards. The demand for geographers also may be dampened by larger trends in the national economy, including the privatization of government jobs, the traditional backbone of geographers' employment; the displacement of aerospace workers with complementary skills; and economic globalization, which carries the potential for substituting foreign for domestic workers. During the 1980s, the discipline did a phenomenal job of attracting new students who came to see geography as the stepping-stone to a satisfying and productive career. Our challenge for the 1990s is in identifying new niches of employment for the students we have so successfully attracted to the field, in marketing ourselves to potential employers as effectively as we have to potential students, and in helping students make the difficult transition to a highly volatile, competitive, and uncertain labor market.

Our interviews with AAG corporate sponsors have implications for the undergraduate curriculum. The debate over geography as a broad-based liberal arts discipline or as a technical, semiprofessional field ignores the realities of the current labor market. Sponsors told us that they want employees who can combine technical skills with a broad-based background. Geography's comparative advantage over other social sciences lies in its ability to combine technical skills with a more traditional liberal arts perspective. Successful geography programs will be those that are able to find the appropriate balance of field-based technical skills like GIS, cartography, and air photo interpretation with competence in literacy, numeracy, decision making, problem solving, and critical thinking.

The effects of geographic education initiatives on the labor market for geographers will be played out over a geographically disparate landscape of teacher certification requirements, high school geography requirements, and university entrance requirements. Because education, including geographic education, is largely a local matter, local geographers are best equipped to keep tabs on state certification requirements and their effects on the demand for geography teachers, university requirements and their effects on precollegiate geography training, and the trickle-up effects of the implementation of National Geography Standards. ■

Literature Cited

Farnham, A. 1993. Out of college, what's next? *Fortune* July 12:58–64.

Fierman, J. 1994. The contingency workforce. *Fortune* January 23:30–36.

Kutscher, R. E. 1993. Historic trends, 1950–1992, and current uncertainties. *Monthly Labor Review* 116(110):3–10.

Martin, J. 1992. Your new global workforce. *Fortune* December 14:52–60.

Swan, P. N. 1992. Early retirement incentives with upper age limits under the Older Workers Benefits Protection Act. *Journal of College and University Law* 19:53–72.

PATRICIA GOBER is Professor of Geography at Arizona State University, Tempe, AZ 85287-0104. Her areas of interest are population and urban geography.

AMY GLASMEIER is Associate Professor of Geography and Regional Planning at Pennsylvania State University, University Park, PA 19802. Her areas of expertise include economic and industrial geography, public policy, and trade and technology.

JAMES M. GOODMAN is Geographer-in-Residence at the National Geographic Society, 1145 17th Street NW, Washington, DC 20036-4688. His areas of interest are geography alliance management, outreach programs in geographic education, and American Indian lands and resources.

DAVID PLANE is Professor and Head of the Department of Geography and Regional Development at the University of Arizona, Tucson, AZ 85721. His areas of interest are population geography and regional science.

HOWARD STAFFORD is Professor of Geography at the University of Cincinnati, Cincinnati, OH 45221-0131. His areas of interest are industrial location and regional economic development.

JOSEPH WOOD is Associate Professor and Chair in the Department of Geography and Earth Systems Science at George Mason University, Fairfax, VA 22030-4444. He studies and writes on the American settlement landscape, from colonial New England villages to contemporary suburbs, and is presently investigating the increasing Vietnamese presence in northern Virginia.

Appendix B

Professional Organizations in U.S. Geography

Geography in the United States is represented by four associations that vary in size and purpose:

American Geographical Society
156 Fifth Avenue, Suite 600
New York, NY 10010
(212) 242-0214
Fax: (212) 989-1583
E-mail: amgeosoc@village.ios.com

Association of American Geographers
1710 Sixteenth Street, NW
Washington, DC 20009
(202) 234-1450
Fax: (202) 234-2744
E-mail: gaia@aag.org

National Council for Geographic Education
16A Leonard Hall
Indiana University of Pennsylvania
Indiana, PA 15705
(412) 357-6290
Fax: (412) 357-7708
E-mail: clmccard@grove.iup.edu

National Geographic Society
1145 Seventeenth Street, NW
Washington, DC 20036
(202) 857-7000
Fax: (202) 775-6141
E-mail: webmaster@ngs.org

Outside the United States, American geographers and their organizations have strong links with the International Geographical Union, the British Royal Geographical Society (Institute of British Geographers), and the Canadian Association of Geographers.

International Geographical Union
Department of Geography
University of Bonn
Meckenheimer Allee 166
Germany
Phone: (49 228) 73 9287
Fax: (49 228) 73 9272
E-mail: secretariat@igu.bn.eunet.de

Royal Geographical Society (with the Institute of British Geographers)
1 Kensington Gore
London, England SW7 2AR
(44 171) 589-5466
Fax: (44 171) 584-4447

Canadian Association of Geographers
Burnside Hall
McGill University
Montreal, Quebec H3A 2K6
Canada
(514) 398-4946
Fax: (514) 398-7437
Email: cag@felix.geog.mcgill.ca

Appendix C

Biographical Sketches of Committee Members

THOMAS J. WILBANKS (*Chair*) is a corporate research fellow at the Oak Ridge National Laboratory, where he leads the laboratory's energy, environmental, and science and technology programs in developing countries. A former president of the Association of American Geographers, his research interests are in processes and mechanisms for realizing sustainable development, energy and environmental policy, and institutional capacity building. He was awarded Honors by the Association of American Geographers in 1986, the Distinguished Geography Educator's Award of the National Geographic Society in 1993, and the Anderson Medal of Honor in Applied Geography in 1995, and he is a fellow of the American Association for the Advancement of Science. He received his M.A. and Ph.D. degrees from Syracuse University.

ROBERT McC. ADAMS, an anthropologist, was secretary of the Smithsonian Institution from 1984 until his retirement in 1994. Previously, he was a member of the University of Chicago faculty, serving at various times as director of its Oriental Institute, dean of social sciences, and provost. Currently, he is adjunct professor of anthropology at the University of California, San Diego. He is a member of the National Academy of Sciences, the American Philosophical Society, and the American Academy of Arts and Sciences.

Originally having specialized and conducted many years of fieldwork in the historical geography and archeology of the Near East, his more recent interests have focused on contexts of innovation and the history of technology. **Paths of Fire: An Anthropologist's Inquiry into Western Technology** was published in October 1996 by Princeton University Press, Princeton, New Jersey. His earlier publications include **Heartland of Cities: Surveys of Ancient Settlement and Land Use on the Central Floodplain of the Euphrates** (University of Chicago Press, Chicago, Illinois, 1981) and **The Evolution of Urban Society: Early Mesopotamia and Prehispanic Mexico** (Aldine Publishing Company, Chicago, Illinois, 1966).

MARTHA E. CHURCH came to Hood College in 1975 as its first woman president. A graduate of Wellesley College, she earned her M.A. degree from the University of Pittsburgh and her Ph.D. from the University of Chicago. She is also the recipient of nine honorary degrees. She retired as Hood's president on June 30, 1995, and assumed a part-time appointment as senior scholar at the Carnegie Foundation for the Advancement of Teaching on July 1, 1995.

In recognition of her career as a geographer and successful college administrator, she was elected in the spring of 1989 to the Board of Trustees of the National Geographic Society and also to the Board of Trustees of the society's Education Foundation. She chairs the society's Audit Review Committee and serves on its Executive, Compensation, and Nominating committees.

WILLIAM A.V. CLARK is professor of geography and chair of the Department of Geography at the University of California, Los Angeles. He received his B.A. and M.A. degrees from the University of New Zealand and his Ph.D. from the University of Illinois at Urbana. He also has a D.Sc. degree from the University of Auckland, New Zealand, and the Doctorem Honoris Causa from the University of Utrecht, The Netherlands. His research interests include analyses of migration and residential mobility and the nature of demographic change in large cities. He is a member of the Association of American Geographers, the Population Association of America, and the New Zealand Geographical Society.

ANTHONY R. DE SOUZA is currently professor of geography at Southwest Texas State University. Previously, he was executive director of the Geography Education Standards Project, secretary general of the 27th International Geographical Union Congress, editor of *National Geographic Research & Exploration*, and Editor of the *Journal of Geography*. He has also held positions as professor and visiting professor of geography at the George Washington University, the University of Wisconsin-Eau Claire, the University of Minnesota, the University of California-Berkeley, and the University of Dar es Salaam in Tanzania. He obtained his B.A. and Ph.D. in geography from the University of Reading, England, and has received numerous honors and awards, including the Gilbert Grosvenor Honors for Geographic Education of the Association of American Geographers in 1996. His teaching and research interests include geography education and regional economic development.

PATRICIA P. GILMARTIN is professor of geography at the University of South Carolina. Previously, she was a member of the geography faculty at the University of Victoria in British Columbia, Canada. Dr. Gilmartin received her master's degree from Georgia State University and her Ph.D. from the University of Kansas. She is a member of the North American Cartographic Information Society, the Canadian Cartographic Association, the Society of Woman Geographers, and the Association of American Geographers, for whom she currently serves as a national councilor and chair of the Publications Committee. Her research interests include map design, cognitive and perceptual processes involved in map use and way finding, and women explorers and travelers.

WILLIAM L. GRAF is regents' professor of geography at Arizona State University. He obtained his Ph.D. from the University of Wisconsin, Madison, with a major in physical geography and a minor in water resources management. His specialties include fluvial geomorphology and policy for public land and water. The focus of much of his geomorphologic research and teaching has been on river channel change, human impacts on river processes and morphology, and contaminant transport and storage in river sediments, especially in dryland rivers. In the area of public policy he has emphasized the interaction of science and decision making and resolution of conflicts between economic development and environmental preservation. His work has been funded by federal, state, and local agencies, ranging from the National Science Foundation, the U.S. Environmental Protection Agency, and the U.S. Army Corps of Engineers to cities, tribes, and private companies. He has published more than 100 papers, articles, book chapters, and reports; his books include **Geomorphic Systems of North America** (Geological Society of America, Boulder, Colorado, 1987), **The Colorado River: Instability and Basin Management** (Association of American Geographers, Washington, D.C., 1985), **Fluvial Processes in Dryland Rivers** (Springer-Verlag, Berlin, New York, 1988), **Wilderness Preservation and the Sagebrush Rebellions** (Rowman & Littlefield, Savage, Maryland, 1990), and **Plutonium and the Rio Grande: Environmental Change and Contamination in the Nuclear Age** (Oxford University Press, New York, 1994). His work has produced awards from the Association of American Geographers and the Geological Society of America, as well as a Guggenheim Fellowship. He has served the National Research Council in numerous capacities, including as a member of its Water and Science Technology Board and the Committee on Glen Canyon Environmental Studies and as chair of the Workshop to Advise the President's Council on Sustainable Development and the Committee on Innovative Watershed Management.

JAMES W. HARRINGTON directs the Geography and Regional Science Program of the National Science Foundation's Division of Social, Behavioral, and Economic Research. He is also an associate professor of public policy and geography at George Mason University. His research area is economic development of subnational regions (e.g., metropolitan areas), specifically the roles of service-sector activity and of international trade. His book **Industrial Location: Theory and Practice** (with Barney Warf) was published by Routledge in 1995. He serves as secretary and councilor of the Association of American Geographers and as executive director of the North American Regional Science Council. He also has served on civic and church boards in Buffalo, New York, and Reston, Virginia. Raised in South Carolina, he holds an A.B. degree from Harvard University and M.A. and Ph.D. degrees from the University of Washington.

SALLY P. HORN is an associate professor of geography at the University of Tennessee, Knoxville. Her research areas are biogeography and paleoecology, especially in the Latin American tropics. She received her B.A., M.A., and Ph.D. (geography) degrees from the University of California, Berkeley. She is a member of the Association of American Geographers, the Ecological Society of America, the Association for Tropical Biology, and the American Quaternary Association.

ROBERT W. KATES is an independent scholar in Trenton, Maine, and university professor (emeritus) at Brown University. He is an executive editor of *Environment* magazine, distinguished scientist at the George Perkins Marsh Institute at Clark University, faculty associate of the College of the Atlantic, and senior fellow at the H. John Heinz III Center for Science, Economics and the Environment. Between 1986–1992 he directed the Alan Shawn Feinstein World Hunger Program at Brown University. Before 1986 he held various teaching and research posts at Clark University and the University of Dar es Salaam. He has received a National Medal of Science (1991), a MacArthur Prize fellowship (1981–1985), an honors award from the Association of American Geographers, and an honorary degree from Clark University (1993). He is a member of the National Academy of Sciences and the American Academy of Arts and Sciences, a foreign member of the Academia Europaea, and a fellow of the American Association for the Advancement of Science. In 1993–1994, he was the president of the Association of American Geographers. He holds a Ph.D. degree in geography from the University of Chicago.

ALAN MacEACHREN is professor of geography at Penn State University. He received his B.A. in geography from Ohio University and his M.A. and Ph.D. in geography from the University of Kansas. Currently, he chairs the U.S. National Committee for the International Cartographic Association (ICA) and the ICA Commission on Visualization and is codirector of the Geographic Information Analysis Core of the Population Research Institute at Penn State. His research interests include the integration of cognitive and semiotic approaches to geoefereneced (particularly cartographic) representation, methods for geo-referenced visualization and data mining, environmental cognition and the potential interaction between geo-information technology and cognition of space (in contexts such as way finding and geographic education), and social consequences of "mapping" (both past and present). He is a member of the Association of American Geographers, the Cartography and Geographic Information Society, the Canadian Cartographic Association, the British Cartographic Society, the Society of Cartographers, the North American Cartographic Information Society, and the National Council on Geographic Education.

ALEXANDER B. MURPHY is professor of geography and head of the Department of Geography at the University of Oregon. He specializes in political, cultural, and ethnic geography, with a regional emphasis on Europe. In 1991 he received a National Science Foundation Presidential Young Investigator Award. He is a councilor of the American Geographical Society and North American editor of *Progress in Human Geography*. He has headed a National Science Foundation study exploring ways that geographers can contribute to a proposed initiative on democratization. He also chairs a task force overseeing the development of an advanced placement course and examination in geography. He holds a bachelor's degree in archeology from Yale University, a law degree from Columbia University School of Law, and a Ph.D. in geography from the University of Chicago.

GERARD RUSHTON is professor of geography at the University of Iowa. He received his B.A. and M.A. degrees from the University of Wales and his Ph.D. in geography from the University of Iowa. He served as a member of the National Research Council's Mapping Science Committee (1990/1993) and was a member of the Board of Directors, National Center for Geographic Information and Analysis (1989/1992). His research areas include spatial analysis of disease, spatial decision support systems, and methods of optimal location of facilities. His recent writings have appeared in *Environment and Planning A, B, & C*, *Statistics in Medicine*, *Socio-Economic Planning Sciences*, and *Papers, Regional Science Association*.

ERIC S. SHEPPARD is professor of geography at the University of Minnesota. He received his M.A. and Ph.D. from the University of Toronto and his B.Sc. (Hons.) from Bristol University and has been a visiting scholar at the International Institute of Applied Systems Analysis (Austria) and at the Universities of London, Vienna, Melbourne, and Indonesia. His research interests include the political and economic dynamics of urban and regional systems, spatial strategies of multiplant firms, spatialization of economic and social theory, and interrelationships between philosophical traditions of geographic thought and the methodologies of geographic analysis. He is a member of the Association of American Geographers, the Canadian Association of Geographers, and the Regional Science Association International.

BILLIE LEE TURNER II is the Higgins Chair of Environment and Society, Graduate School of Geography, and director, George Perkins Marsh Institute, Clark University. His research interests are human-environment relationships and range from ancient agriculture and environment in Mexico and Central America to contemporary agricultural change in the tropics and global land-use change. He received his B.A. and M.A. degrees from the University of Texas and his Ph.D. from the University of Wisconsin, Madison. He is a member of the National Academy of Sciences.

CORT J. WILLMOTT received his Ph.D. from UCLA in 1977. He is currently a professor of geography and marine studies at the University of Delaware, where he also serves as chair of the Department of Geography, director of the University's Center for Climatic Research, director of the Environmental Science Program of the College of Arts and Science, and associate director of Delaware's space grant program. His research interests include large-scale climate variability and change, land surface processes and their influences on climate, spatial interpolation over extensive geographic domains, and statistical evaluation of model performance. Grants primarily from the National Aeronautics and Space Administration and the National Science Foundation have supported Professor Willmott's research since 1980. He is a member of the Association of American Geographers, the American Meteorological Society, the American Geophysical Union, and Sigma Xi.

Index

D

E

H

I

V

W

Y